EXTREME WEATHER

EXTREME WEATHER

A GUIDE TO SURVIVING FLASH FLOODS,
TORNADOES, HURRICANES, HEAT
WAVES, SNOWSTORMS, TSUNAMIS,
AND OTHER NATURAL DISASTERS

BONNIE SCHNEIDER

FOREWORD BY MAX MAYFIELD

palgrave
macmillan

EXTREME WEATHER
Copyright © Bonnie Schneider, 2012.
All rights reserved.

First published in 2012 by PALGRAVE MACMILLAN® in the United
States—a division of St. Martin's Press LLC, 175 Fifth Avenue, New
York, NY 10010.

Where this book is distributed in the UK, Europe and the rest of the
world, this is by Palgrave Macmillan, a division of Macmillan Publishers
Limited, registered in England, company number 785998, of Houndmills,
Basingstoke, Hampshire RG21 6XS.

Palgrave Macmillan is the global academic imprint of the above
companies and has companies and representatives throughout the world.

Palgrave® and Macmillan® are registered trademarks in the United
States, the United Kingdom, Europe and other countries.

ISBN 978-0-230-11573-6

Library of Congress Cataloging-in-Publication Data

Schneider, Bonnie.
    Extreme weather : a guide to surviving flash floods, tornadoes,
hurricanes, heat waves, snowstorms, tsunamis, and other natural
disasters / Bonnie Schneider ; foreword by Max Mayfield.
        p.   cm.
    Includes index.
    ISBN 978-0-230-11573-6 (pbk.)
    1. Emergency management. 2. Survival. 3. Safety
education. I. Title.
HV551.2.S393   2012
613.6'9—dc23

                                                            2011026831

A catalogue record of the book is available from the British Library.

Design by Letra Libre, Inc.

First edition: January 2012

10 9 8 7 6 5 4 3 2 1

Printed in the United States of America.

CONTENTS

FOREWORD

IT SEEMS LIKE I HAVE EITHER BEEN PREPARING FOR A
storm, going through a storm, or recovering from a storm for the
past 40 years. I have personally observed the damage caused by
hurricanes during my nearly 35 years of working at the National
Hurricane Center (NHC) followed by five years as a Hurricane
Specialist for WPLG-TV. While I was Director of the NHC from
2000 to 2007, I would try to visit the most impacted areas after major US hurricane landfalls. One of the reasons for doing
this was to get firsthand accounts from emergency managers and
other public officials and people who had experienced the hurricane to help identify areas needing improvement in the US hurricane program. There are always lessons to be learned from past
hurricanes, and this was especially true during the extremely active 2004 and 2005 hurricane seasons.

One thing that stood out to me was that people who had a
hurricane plan and could execute that plan nearly always did better than people who did not have a plan. There is always a tre-
mendous amount of stress involved during any hurricane threat,
and this stress can be decreased considerably if people are prepared and know what to do.

Being prepared and knowing what to do also apply to other
natural disasters as well. It helps to understand the basic science
behind hurricanes, tornadoes, floods, extreme heat, wildfires, rip
currents, landslides, tsunamis, severe thunderstorms, snow and

ice storms, and earthquakes. Bonnie Schneider shares some unforgettable stories on each of these phenomena in *Extreme Weather* that will motivate you to prepare in case you find yourself in life-threatening situations similar to those described in the stories.

Extreme Weather will educate you on the science behind the weather and earthquakes. And more importantly, the book gives tips from FEMA, NOAA, and other reliable sources to guide you in preparing a family disaster plan for people and for pets. Loss of life is preventable in most cases.

In addition to providing disaster planning tips from government agencies like FEMA and NOAA, Ms. Schneider also interviewed a wide variety of experts. These include an arctic cold weather survival expert from the US Air Force in Alaska, an assistant fire chief in Boulder, Colorado, an expert with the Los Angeles County Fire Department, a captain in a SWIFT Water Rescue team in Tennessee, the head of the United States Life Saving Association, earthquake and tsunami experts from universities, and pet experts across the United States and abroad. When you read their advice in *Extreme Weather,* it is clear that these experts don't just talk the talk. They have walked the walk. Their specialized vantage points offer readers a firsthand perspective in addition to valuable safety and preparation advice.

I often think of philosopher George Santayana's quote: "Those who cannot remember the past are condemned to repeat it." Let's learn from the past mistakes of others. Heeding the advice in this book could make you a survivor rather than a victim. It could save your life.

—*Max Mayfield, Hurricane Specialist for*
WPLG-TV, Miami, Florida, and
former Director of the National Hurricane Center

INTRODUCTION

THROUGHOUT THE MANY YEARS I WORKED IN LOCAL TV news, I had the privilege to get out into the community and report the weather from various festivals, fairs, and sporting events. Many of those TV segments were broadcast on nice, sunny days. However, there were other times when the weather was quite different. I've stood on the shoulder of the Long Island Expressway for hours of live reports during brutal nor'easters with ice pellets stinging my face, my feet numb in knee-deep snow. I braced myself to stand upright in strong winds along coastal areas of Louisiana, New York, and Massachusetts in anticipation of an approaching tropical storm or hurricane. Each event was intense and memorable.

However, it wasn't until I started working for CNN in 2005 that I began to report on the largest and most dangerous weather events that affected millions of people around the world. Breaking news was happening daily, and to be in the studio while the information was coming in made it more crucial to get accurate updates out as quickly as possible. One evening in August 2005, I was driving to work to do overnight weather reports in the hours before Hurricane Katrina made landfall on the Gulf Coast. In my mind, I kept replaying the severity of the warnings that I had read that day and evening. I was worried—I wondered if people in Louisiana and Mississippi realized how bad this was going to be. Were they prepared?

In the years that followed, more extreme weather events and natural disasters occurred with increasing frequency. "Some 373 natural disasters killed over 296,800 people in 2010, costing nearly $110 billion, according to the Centre for Research on the Epidemiology of Disasters (CRED). This made 2010 the deadliest year for natural disasters in at least two decades."[1] In 2011, record-breaking tornadoes wiped out communities in Alabama and Missouri. The Mississippi River rose to dangerous flood levels. Wildfires burned with fury through Texas, Arizona, and New Mexico.

The news media's headlines were filled with epic stories of fear and survival in these devastating natural disasters. This was reflected across social media as well. Trending topics like #wildfire, #earthquake, and #tsunami became more widely used. On my social media pages on Facebook and Twitter (@BonnieWeather), I began to receive more and more questions from viewers about extreme weather:

- "What causes a tornado to form?"
- "What should you do if you are under a tsunami warning?"
- "How can you swim safely to shore if you are caught in a rip current?"
- "What emergency supplies should I keep at home?"
- "How can I protect my pet if I have to evacuate?"

This is how *Extreme Weather* was born. I wanted to write a book that was well researched, easy to understand, and would serve readers as a comprehensive resource to help keep them and their families—including their pets—safe.

I hope that the stories of survival in this book will not only inspire you, but that they will also demonstrate how quickly a person's life can change forever. My goal is that the book will help you be more proactive in preparing for extreme weather. If tragedy does strike, this book can also give you some important steps to take in the aftermath of an extreme weather event or a natural disaster.

—*Bonnie Schneider*
Atlanta 2011

1
HURRICANES

At 2:30 a.m. on Monday, August 29, 2005, the blast of the telephone ringing woke 47-year-old Mary Theriot of Chalmette, Louisiana, from a deep sleep.[1]

"Hello?"

"Mary! It's Rose. I'm in Mobile, watching this storm on TV, and it's gettin' real bad! They're sayin' it's headed straight for you in N'awlins."

"Are you sure?"

"Yes! Whatever you do, don't go back to sleep. Wake up! Are you ready if it hits?"

"Yes, hang on Rose," Mary told her friend. "I'm up. I'm going to take the phone downstairs with me and make some coffee. Don't worry, I'm not going back to sleep."

They talked a little longer, and then the phone abruptly cut off at 3 a.m. The lines were dead. Power was out.

For hours, Mary watched in the dark outside her kitchen window. Her husband, Joe, and her 11-year-old daughter, Cheyenne, were asleep.

The wind was getting stronger. It was howling loudly through the glass. Large trees had been completely uprooted and were lying in the yard. Some were on top of cars. At around 8 a.m., there was enough light to go out back and check on things. Trash and lawn chairs were everywhere. Even though they had no power, it wasn't so bad, Mary thought. Things could have been worse.

Looking past downed trees and power lines, Mary noticed an old pickup truck in her neighbor's yard that was now drenched in muddy water. The water level was halfway up the tire. She surveyed the damage to other neighbors' homes on her street.

Then Mary brought her gaze back to that pickup. It had been only a few minutes, but now the water covered all four tires and was almost at the hood of the car.

"Cheyenne, get Joe!" Mary called to her daughter. "We are fixin' to flood and flood bad!"

Mary's heart was racing. Once inside the house, the three of them frantically started picking things up off the floor. Pictures, books, and anything they could grab went on high shelves or on top of the refrigerator. Water began to seep in from under the walls. Mary looked out the front door; the porch they'd been standing on moments ago was no longer visible. All she could see was that brown, oily water. Mary felt herself panicking.

"Hurry!" she yelled. They worked faster gathering what they could. Cheyenne climbed up the ladder to the attic. Mary grabbed her cat Sasha and handed her up to Cheyenne. But where was Otis? Mary had rescued both cats and considered them her babies. Otis was hiding in the closet. Mary scooped him up, her hands squeezing his wet paws.

"Bless his heart! He's scared out of his mind!"

By Wednesday morning, the water had receded enough for officials to drive into their neighborhood. A local police officer arrived to announce a mandatory evacuation. He told Mary that her family pets were not allowed at the shelter unless they were very small and caged. The cage for Mary's two cats was too large to carry. She'd have to leave Sasha and Otis behind. Before leaving, Mary set out plenty of food and water for her beloved cats. Joe knocked out a window screen so the animals could get out if they had to. Mary had raised Sasha and Otis from the time they were kittens, and she felt like she was leaving part of her family behind.

The weather was hot and humid. Carrying what they could of their belongings in garbage bags, the family eventually found their way to a shelter. Days later, they finally made contact with Joe's relatives. They then went to Joe's family home in Assumption Parish, where they could rest and recover. It would be months before they would return home to Chalmette.

When Hurricane Katrina made landfall on the Gulf Coast, it came in as a Category 3 storm, with the eye passing east of New Orleans. This meant the strongest winds, those found in the northeastern quadrant, didn't hit Louisiana. Mississippi saw much worse wind damage.

One of the biggest problems was the overtopping and breaching of the levees. In New Orleans, there were at least 50 levee failures after Katrina hit: water overtopped some levees completely; others just broke down. The failure of the levees released billions of gallons of water into the city and surrounding areas.

In Mary Theriot's town of Chalmette, only about a mile east of the Lower Ninth Ward, a 15- to 19-foot surge of water flooded the town through the Mississippi River Gulf Outlet. Chalmette was completely underwater and remained that way well after the storm had passed. Many of the homes and buildings were permanently destroyed.

Mary and her family lived with relatives for two months. Wondering what had become of the cats, Mary went back home the first time briefly in September, and found her cat Sasha, but not Otis. In November, the Theriot family set out to Chalmette for the second time since Katrina. On this trip back to Chalmette, Mary had given up on seeing Otis again. She didn't even bother taking his cage with her this time.

The Theriot family drove into their old neighborhood in Chalmette. They were turning off Delille Street and making a left onto West Moreau when they saw him. The cat was stretched out on the sunny driveway of the home of one of their old neighbors. Mary swung the passenger door open before Joe could even hit the brakes.

"Otis!" she screamed, rushing toward him. "Otis! My God! There you are!"

Joe screeched the car to a halt. Cheyenne jumped out. Otis was alive! He was safe. They were all safe. They lost their possessions, lost their home, but they were safe. They survived Hurricane Katrina.

—*Hurricane Katrina, August 29, 2005. Courtesy of the National Guard*

Over 1,800 people lost their lives to Hurricane Katrina. It was the costliest storm ever, with more than $80 billion in property damage. The summer of 2005 was the most active year on record by almost all standards. In total there were a record 15 hurricanes, 7 of them major ones (meaning Category 3 or stronger, sustained winds greater than 110 mph), and a record-breaking 4 of the storms contained sustained wind speeds that climbed to greater than 155 mph.

AS HURRICANES DEVELOP FROM RELATIVELY SMALL clusters of thunderstorms, they may initially be categorized as tropical depressions, with sustained wind speeds of 38 mph or less, or tropical storms, with sustained wind speeds of 39 to 73 mph. These events might not boast the extreme wind speeds that hurricanes do, but don't write them off. Tropical depressions and tropical storms can do plenty of damage from flooding and wind, particularly if they are slow-moving. The longer a storm lingers over a given area, the more rain it can produce and the higher the likelihood that the region will experience flooding.

Though you might think the monumental force of 100 mph+ winds would be the most deadly effect of hurricanes, **flooding is the leading killer in hurricanes.** That is particularly true on the coast, where storm surge, an abnormal rise in water generated by a storm, can occur.

Even if you don't live or vacation directly in a coastal area, hurricanes can still affect you. Strong hurricanes can cause flooding well inland from where the storm makes landfall. They also can spawn tornadoes and severe thunderstorms far from the coast.

The National Hurricane Center issues the advisories on tropical cyclones in the Atlantic and East Pacific Basins. Understanding the difference between these advisories could save your life. Note that a *hurricane warning* poses more of an immediate threat than a *hurricane watch*. Because hurricane preparedness activities become difficult once winds reach tropical storm force, a hurricane watch is issued 48 hours in advance of the anticipated onset of these dangerous winds. A hurricane warning is issued 36 hours in advance.

Hurricane Watch: An announcement that sustained winds of 74 mph or higher are possible within the specified area.

Hurricane Warning: An announcement that sustained winds of 74 mph or higher are expected within the specified area.

Tropical Storm Watch: An announcement that sustained winds of 39 to 73 mph are possible within the specified area within 48 hours.

Tropical Storm Warning: An announcement that sustained winds of 39 to 73 mph are expected somewhere within the specified coastal area within 36 hours.

THE SAFFIR-SIMPSON HURRICANE WIND SCALE

In order to be fully prepared, it's advisable to learn the way public officials classify hurricanes and the potential damage they can bring. The Saffir-Simpson Hurricane Wind Scale[2] evaluates hurricane winds and their projected damage using a 1 to 5 categorization based on the hurricane's intensity at the indicated time. The Saffir-Simpson Scale has been used to alert the public to landfalling hurricanes since 1973. It was updated in 2010 to include more specific information. In general, damage rises by a factor of four for every category increase.

CATEGORY 1 HURRICANE	*Winds (1 min. sustained winds in mph and km/hr)* *74–95 mph; 64–82 kt; 119–153 km/hr*
Summary	Very dangerous winds will produce some damage
People, Livestock, and Pets	People, livestock, and pets struck by flying or falling debris could be injured or killed.
Mobile Homes	Older (mainly pre-1994 construction) mobile homes could be destroyed, especially if they are not anchored properly as they tend to shift or roll off their foundations. Newer mobile homes that are anchored properly can sustain damage involving the removal of shingle or metal roof coverings and loss of vinyl siding, as well as damage to carports, sunrooms, or lanais.
Frame Homes	Some poorly constructed frame homes can experience major damage, involving loss of the roof covering and damage to gable ends as well as the removal of porch coverings and awnings. Unprotected windows may break if struck by flying debris. Masonry chimneys can be toppled. Well-constructed frame homes could have damage to roof shingles, vinyl siding, soffit panels, and gutters. Failure of aluminum, screened-in, swimming pool enclosures can occur.
Apartments, Shopping Centers, and Industrial Buildings	Some apartment building and shopping center roof coverings could be partially removed. Industrial buildings can lose roofing and siding especially from windward corners, rakes, and eaves. Failures to overhead doors and unprotected windows will be common.
High-Rise Windows and Glass	Windows in high-rise buildings can be broken by flying debris. Falling and broken glass will pose a significant danger even after the storm.
Signage, Fences, and Canopies	There will be occasional damage to commercial signage, fences, and canopies.
Trees	Large branches of trees will snap and shallow-rooted trees can be toppled.
Power and Water	Extensive damage to power lines and poles will likely result in power outages that could last a few to several days.
Example	Hurricane Dolly (2008) is an example of a hurricane that brought Category 1 winds and impacts to South Padre Island, Texas.

CATEGORY 2 HURRICANE

Winds *(1 min sustained winds in mph and km/hr)*
96–110 mph; 83–95 kt; 154–177 km/hr

Summary	Extremely dangerous winds will cause extensive damage
People, Livestock, and Pets	There is a substantial risk of injury or death to people, livestock, and pets due to flying and falling debris.
Mobile Homes	Older (mainly pre-1994 construction) mobile homes have a very high chance of being destroyed, and the flying debris generated can shred nearby mobile homes. Newer mobile homes can also be destroyed.
Frame Homes	Poorly constructed frame homes have a high chance of having their roof structures removed, especially if they are not anchored properly. Unprotected windows will have a high probability of being broken by flying debris. Well-constructed frame homes could sustain major roof and siding damage. Failure of aluminum, screened-in, swimming pool enclosures will be common.
Apartments, Shopping Centers, and Industrial Buildings	There will be a substantial percentage of roof and siding damage to apartment buildings and industrial buildings. Unreinforced masonry walls can collapse.
High-Rise Windows and Glass	Windows in high-rise buildings can be broken by flying debris. Falling and broken glass will pose a significant danger even after the storm.
Signage, Fences, and Canopies	Commercial signage, fences, and canopies will be damaged and often destroyed.
Trees	Many shallowly rooted trees will be snapped or uprooted and block numerous roads.
Power and Water	Near-total power loss is expected with outages that could last from several days to weeks. Potable water could become scarce as filtration systems begin to fail.
Example	Hurricane Frances (2004) is an example of a hurricane that brought Category 2 winds and impacts to coastal portions of Port St. Lucie, Florida, with Category 1 conditions experienced elsewhere in the city.

CATEGORY 3 HURRICANE

Winds (1 min sustained winds in mph and km/hr)
111–130 mph; 96–113 kt; 178–209 km/hr

Summary	Devastating damage will occur
People, Livestock, and Pets	There is a high risk of injury or death to people, livestock, and pets due to flying and falling debris.
Mobile Homes	Nearly all older (pre-1994) mobile homes will be destroyed. Most newer mobile homes will sustain severe damage with potential for complete roof failure and wall collapse.
Frame Homes	Poorly constructed frame homes can be destroyed by the removal of the roof and exterior walls. Unprotected windows will be broken by flying debris. Well-built frame homes can experience major damage involving the removal of roof decking and gable ends.
Apartments, Shopping Centers, and Industrial Buildings	There will be a high percentage of roof covering and siding damage to apartment buildings and industrial buildings. Isolated structural damage to wood or steel framing can occur. Complete failure of older metal buildings is possible, and older unreinforced masonry buildings can collapse.
High-Rise Windows and Glass	Numerous windows will be blown out of high-rise buildings resulting in falling glass, which will pose a threat for days to weeks after the storm.
Signage, Fences, and Canopies	Most commercial signage, fences, and canopies will be destroyed.
Trees	Many trees will be snapped or uprooted, blocking numerous roads.
Power and Water	Electricity and water will be unavailable for several days to a few weeks after the storm passes.
Example	Hurricane Ivan (2004) is an example of a hurricane that brought Category 3 winds and impacts to coastal portions of Gulf Shores, Alabama, with Category 2 conditions experienced elsewhere in this city.

CATEGORY 4 HURRICANE	Winds (1 min sustained winds in mph and km/hr) 131–155 mph; 114–135 kt; 210–249 km/hr
Summary	Catastrophic damage will occur
People, Livestock, and Pets	There is a very high risk of injury or death to people, livestock, and pets due to flying and falling debris.
Mobile Homes	Nearly all older (pre-1994) mobile homes will be destroyed. A high percentage of newer mobile homes also will be destroyed.
Frame Homes	Poorly constructed homes can sustain complete collapse of all walls as well as the loss of the roof structure. Well-built homes also can sustain severe damage with loss of most of the roof structure and/or some exterior walls. Extensive damage to roof coverings, windows, and doors will occur. Large amounts of windborne debris will be lofted into the air. Windborne debris damage will break most unprotected windows and penetrate some protected windows.
Apartments, Shopping Centers, and Industrial Buildings	There will be a high percentage of structural damage to the top floors of apartment buildings. Steel frames in older industrial buildings can collapse. There will be a high percentage of collapse to older unreinforced masonry buildings.
High-Rise Windows and Glass	Most windows will be blown out of high-rise buildings resulting in falling glass, which will pose a threat for days to weeks after the storm.
Signage, Fences, and Canopies	Nearly all commercial signage, fences, and canopies will be destroyed.
Trees	Most trees will be snapped or uprooted and power poles downed. Fallen trees and power poles will isolate residential areas.
Power and Water	Power outages will last for weeks to possibly months. Long-term water shortages will increase human suffering. Most of the area will be uninhabitable for weeks or months.
Example	Hurricane Charley (2004) is an example of a hurricane that brought Category 4 winds and impacts to coastal portions of Punta Gorda, Florida, with Category 3 conditions experienced elsewhere in the city.

CATEGORY 5 HURRICANE	Winds (1 min sustained winds in mph and km/hr) > 155 mph; > 135 kt; > 249 km/hr
Summary	Catastrophic damage will occur
People, Livestock, and Pets	People, livestock, and pets are at very high risk of injury or death from flying or falling debris, even if indoors in mobile homes or framed homes.
Mobile Homes	Almost complete destruction of all mobile homes will occur, regardless of age or construction.
Frame Homes	A high percentage of frame homes will be destroyed, with total roof failure and wall collapse. Extensive damage to roof covers, windows, and doors will occur. Large amounts of windborne debris will be lofted into the air. Windborne debris damage will occur to nearly all unprotected windows and many protected windows.
Apartments, Shopping Centers, and Industrial Buildings	Significant damage to wood roof commercial buildings will occur due to loss of roof sheathing. Complete collapse of many older metal buildings can occur. Most unreinforced masonry walls will fail which can lead to the collapse of the buildings. A high percentage of industrial buildings and low-rise apartment buildings will be destroyed.
High-Rise Windows and Glass	Nearly all windows will be blown out of high-rise buildings resulting in falling glass, which will pose a threat for days to weeks after the storm.
Signage, Fences, and Canopies	Nearly all commercial signage, fences, and canopies will be destroyed.
Trees	Nearly all trees will be snapped or uprooted and power poles downed. Fallen trees and power poles will isolate residential areas.
Power and Water	Power outages will last for weeks to possibly months. Long-term water shortages will increase human suffering. Most of the area will be uninhabitable for weeks or months.
Example	Hurricane Andrew (1992) is an example of a hurricane that brought Category 5 winds and impacts to coastal portions of Cutler Ridge, Florida, with Category 4 conditions experienced elsewhere in south Miami-Dade County.

If you travel internationally, you'll know that hurricanes do occur around the world, though they are called by different names. Note, however, that no matter how they are referred to, they are all the same violent weather events.

- **"hurricane"** (the North Atlantic Ocean, the Northeast Pacific Ocean east of the dateline, or the South Pacific Ocean east of 160E)
- **"typhoon"** (the Northwest Pacific Ocean west of the dateline)
- **"severe tropical cyclone"** (the Southwest Pacific Ocean west of 160E or Southeast Indian Ocean east of 90E)
- **"very severe cyclonic storm"** (the North Indian Ocean)
- **"tropical cyclone"** (the Southwest Indian Ocean)

—Source: NOAA (National Oceanic and Atmospheric Administration)

WHEN DO HURRICANES OCCUR?

The time of year when hurricanes are most likely to occur varies depending on where you are in the world. Hurricanes tend to occur in the summer and early fall months, when ocean waters are warm enough to fuel monster storms. Hurricane season for the Atlantic Basin and the Central Pacific Basin runs from June 1 to November 30. For the Eastern Pacific Basin, the season runs from May 15 to November 30. The Atlantic Basin includes the Atlantic Ocean, Carribean Sea, and Gulf of Mexico. The Eastern Pacific Basin extends to longitude 140° W.

 If you see the devastation after a hurricane, and talk to some of the people who've experienced it, I guarantee that would be motivation to create your own individual hurricane plan.

—Max Mayfield, former director of the
National Hurricane Center in Miami

Precautions for hurricane season should be taken well before June 1, according to agencies like FEMA (Federal Emergency Management Agency) and NOAA.

HOW TO PREPARE YOUR HOME
FOR HURRICANE SEASON[3]

It's advised that you prepare before for hurricane season before it starts. You should have your Family Disaster Plan and Emergency Disaster Supply Kit ready and accessible. (Details on how to make a plan and build a kit are found in Chapter 13 of this book.)

- Have your NOAA weather radio powered up and set to alert mode during the threat of any severe weather; have it on even while you sleep. Most weather radios have the ability to "wake up" and alert you when an advisory has been issued for your area.
- Know your area. Familiarize yourself with your home's location in terms of vulnerability to storm surge and wind, especially with routes you can take if you have to evacuate.
- Secure your property. Clear the outside of any potential debris. Secure windows with permanent storm shutters. Masking tape does *not* prevent windows from breaking.
- Install straps or additional clips to securely fasten your roof to the frame structure. This will reduce roof damage.
- Clear loose and clogged rain gutters and downspout and turn off propane tanks.
- Fill empty bottles with water for drinking. Disinfect your bathtub and fill it with water for bathing or to flush the toilet if you have to.
- Make sure all family members know how to turn off water, gas, and electricity and how to dial 911.

- (More information on food and water storage supplies can be found in Chapter 14.)
- Strengthen garage doors. Hurricane winds can enter through a damaged garage door, lift the roof, and destroy the home.
- Keep your car filled with gas in case you have to leave suddenly.

WHAT TO DO IF YOU LIVE IN A MOBILE HOME

- Before a storm, check tie-downs for rust and breakage. Make sure they are as secure as possible.
- **Mobile homes can be destroyed or sustain serious damage even in a Category 1 Hurricane** (sustained winds 74 to 110 mph) This is particularly true for mobile homes constructed before 1994. These homes are not completely safe in even minimal-strength hurricanes or tropical storms.
- Evacuate as soon as possible when told to do so by local authorities.

WHAT TO DO IF YOU LIVE IN A HIGH-RISE APARTMENT OR CONDO

- Keep in mind that winds are stronger at higher elevations.
- Condo owners should consider investing in storm shutters or impact-resistant glass for doors and windows.
- Make sure all exits are clearly marked in the hallways.
- Memorize where the stairs are in case the hall goes dark from loss of power.
- Never take the elevator in an emergency situation; use the stairs.
- If you have a balcony or terrace, secure all items on it. They can become airborne and damage your apartment or your neighbors'.
- Make sure your renter's or condo's insurance policy is up-to-date.

WHAT TO DO IF YOU ARE HOME WHEN A HURRICANE STRIKES

- Have your NOAA radio close, along with a supply of batteries, turned on for an update every 15 to 30 minutes. Hurricanes can take hours to pass, and your weather radio is your best source of information during the storm.
- Stay indoors during the hurricane and away from windows, skylights, and glass doors.
- Close all interior doors—secure and brace external doors.
- Take refuge in a small interior room, closet, or hallway on the lowest level. Make sure your Emergency Disaster Supply Kit (contents detailed in Chapter 14) is easily accessible.
- Lie on the floor under a table or another sturdy object.
- Do not be tempted to look out the window or walk outside during the storm to see what's going on; you could be injured by flying debris.
- At some point, you may hear the wind quiet down. This may be the "eye of the hurricane" passing over. The brief, calm period is followed by the back side of the "eye wall," which also contains extremely high winds.
- Do not use any electronic devices, including your phone or computer. A NOAA Weather Radio should be used, as it's battery operated.

IMMEDIATE ACTIONS TO TAKE AFTER THE STORM

- Wait until you hear official word (over your NOAA radio) that the storm is over.
- If you are returning after an evacuation, make sure the building is structurally safe to enter before you go inside.
- Survey the property; be mindful of downed trees and wires.
- Check for exterior structural damage.

Before you go inside, FEMA advises the following precautions. Do not enter if:

- You smell gas.
- Floodwaters remain around the building.
- Your home was damaged by fire and the authorities have not declared it safe.

When you go inside your home, there are certain things you should and should not do. Enter the home carefully and check for damage. Be aware of loose boards and slippery floors. The following items are other things to check inside your home:

- **Natural gas.** If you smell gas or hear a hissing or blowing sound, open a window and leave immediately. If you can, turn off the main gas valve from the outside. Call the gas company from a neighbor's residence. If you shut off the gas supply at the main valve, you will need a professional to turn it back on. Do not smoke or use oil, gas lanterns, candles, or torches for lighting inside a damaged home until you are sure there is no leaking gas or other flammable materials present.
- **Sparks or broken or frayed wires.** Check the electrical system unless you are wet, standing in water, or unsure of your safety. If possible, turn off the electricity at the main fuse box or circuit breaker. If the situation is unsafe, leave the building and call for help. Do not turn on the lights until you are sure they're safe to use. You may want to have an electrician inspect your wiring.
- **Roof, foundation, and chimney cracks.** If it looks like the building may collapse, leave immediately.
- **Appliances.** If appliances are wet, turn off the electricity at the main fuse box or circuit breaker. Then unplug appliances and let them dry out. Have appliances checked by a

professional before using them again. Also have the electrical system checked by an electrician before turning the power back on.

- **Water and sewage systems.** If pipes are damaged, turn off the main water valve. Check with local authorities before using any water; the water could be contaminated. Pump out wells and have the water tested by authorities before drinking. Do not flush toilets until you know that sewage lines are intact.
- **Food and other supplies.** Throw out all food and other supplies that you suspect may have become contaminated or have come into contact with floodwater.
- **Your basement.** If your basement has flooded, pump it out gradually (about one-third of the water per day) to avoid damage. The walls may collapse and the floor may buckle if the basement is pumped out while the surrounding ground is still waterlogged.
- **Open cabinets.** Be alert for objects that may fall.
- **Clean up household chemical spills.** Disinfect items that may have been contaminated by raw sewage, bacteria, or chemicals. Also clean salvageable items.
- **Call your insurance agent.** Keep good records of repair and cleaning costs.

If there is no light when you return, use a flashlight, not candles, to see what you are doing. Once you can see well and begin the cleanup process, take pictures of all the damage to your home and belongings for insurance purposes.

HOW TO PROTECT YOUR PETS WHEN A HURRICANE HITS

Hundreds of animals lose their lives in hurricanes every year. If you're a pet owner, be sure to be prepared as to how you would take care of your pet in the wake of a life-threatening storm, before hurricane season starts.

The first thing to do is make sure your Emergency Disaster Pet Supply Kit is stocked and ready. (Details on how to prepare this kit, with a complete list of items to include, can be found in Chapter 10.)

Dr. Charlotte Krugler, an emergency preparedness veterinarian at Clemson University Livestock Poultry Health, in Columbia, South Carolina, has experience dealing with animals affected by a hurricane. She treated injured animals during Hurricane Hugo in 1989. "Any change from their normal environment could be a stressor for pets," Dr. Krugler says.[4]

Changes like being in a kennel or a motel in the case of an evacuation or being caught outside in the storm can affect animal behavior. She adds that many animals are lost in storms:

> **Make sure each pet is identified or wearing identification. A microchip is the best insurance against getting your pet lost. Keep a photo of yourself with your animal. Also make sure you can put together a detailed description of your pet. Have a "buddy" system with a neighbor to share keys and instructions about each other's pets. Develop and maintain a list, including contact information, of possible temporary shelter sites away from the affected area (friends, motels, boarding kennels).**
>
> **—Dr. Charlotte Krugler, emergency preparedness veterinarian**

If you do have to evacuate, do not leave your animal behind, especially tied to something. Animals can be in more danger if they can't escape flooding or flying debris. Also, if the roads aren't cleared after a storm, you may not be able to return to check on them for days or even weeks.

AFTER THE STORM

After a hurricane hits, your animal may be injured or even traumatized. Dr. Krugler emphasizes it is important to have a properly trained professional animal-behavior expert treat your pet:

  ❝ Stressed animals respond well to care and treatment provided by trained and experienced animal handlers who understand that animal behavior may change in difficult situations.

She adds that, immediately after the storm, the pet care you receive may not be enough to fully treat your animal.

  ❝ Any veterinary treatment provided during or following a hurricane or other disaster even at or near the scene will be basic care such as immobilizing a fracture, cleaning and bandaging wounds, and providing fluids for dehydration. Animals needing more advanced treatment or having chronic illnesses would be referred to a veterinary hospital for care.

CONCLUSION

Hurricanes can strike swiftly and wreak vast destruction. In order to be prepared, know your region's likely susceptibility to hurricanes and tropical storms. Before hurricane season starts, assess your home's ability to withstand the force of a hurricane and secure the house in case the worst occurs. Prepare evacuation plans (if appropriate) and practice them with your family. During hurricane season, keep an eye on storms brewing in the tropics. Hurricane tracks and forecasts are provided free by the National Hurricane Center. Your local media can update you on the storms' projected path, size, intensity, uncertainties, and impacts, as well as provide crucial life-saving information from emergency managers and other local officials. Note, however, that these monster storms can develop, grow, and intensify in just a matter of days. Advance hurricane preparation for your family and pets is paramount for your safety.

2
TORNADOES

"Baby, pray with me!"[1]

Fifty-four-year-old Essie Hendrix pleaded into her cell phone to her husband. She was crunched under a desk in disbelief.

"The storm has hit!"

"Lord . . . have mercy, and save them," her husband, who also happened to be a Baptist minister, answered back from the other end of the phone. His voice was resonant.

"Please, keep them," he implored.

The couple prayed together.

Essie Hendrix was a store manager at Peebles Department Store in Yazoo City, Mississippi. It was a ground-level store that was well-known throughout Yazoo County. Folks came in to shop for everything from clothing to shoes to jewelry and even candles and picture frames. Essie knew many of the customers by name, and they knew her. To them she was "Miss Essie."

The morning of April 24, 2010, started like any other Saturday for Essie. She got to the store at 8:30 a.m., counted down the cash registers, straightened up, and had the doors open for customers promptly at 9 a.m.

"Good morning, Miss Essie!" an older gentleman greeted her as he came in. "You done hear 'bout that storm?" he asked.

"Yes, yes, I have," she told him. "Just looks cloudy now." Essie peered through the glass doors onto North Jerry Clower Boulevard.

But by 10 a.m. the skies had turned dark outside. Essie was getting concerned.

"I thought we might lose power, so I went next door to the supermarket to buy batteries for our flashlights," she recalled.

Essie purchased the batteries and went back inside Peebles. She handed the batteries to the three salesgirls and told them each to get a flashlight. Normally, cell phones were not permitted on the floor. Essie thought each woman should have her phone handy today, though, just in case the power went out.

The sales staff went back to work assisting customers. It was business as usual.

From under the front desk, Essie pulled out her NOAA Weather Radio and turned it on.

"The meteorologist was saying the storm is coming to Yazoo at twelve noon," *she remembers.*

The shopping continued, but everyone around kept talking about the weather. They all had something to say on the subject.

"We'll probably get rain and wind," *most folks agreed.*

"The power would go out," *others offered.*

"But a tornado? Here? In Yazoo City?" *Essie was not convinced.* "It hadn't happened here before. Not to my knowledge."

Throughout the latter part of the hour after 11 a.m., the meteorologist's voice on the radio was getting more intense and urgent.

"He said the storm was coming up from a line near Holly Bluff, up Highway 3."

In a split second Essie had to make a decision.

Essie talks faster as she recalls what she heard next.

"He said the storm was five miles away. It was a bad tornado. In five minutes it was coming . . . to Yazoo. He said, 'Take cover in a sturdy building!'"

At Peebles Department Store in Yazoo City, Essie took charge. She told the sales team: "Close the registers! I need everyone to head to the back of the store right now!" *The woman who was paying for clothes at a nearby register didn't seem to hear her.*

"Ma'am," *Essie said.* "This register is closed. I need you to go to the back of the store. Please, y'all. I need you to go back." *The woman listened and began to walk in the direction Essie indicated.*

"Don't walk . . . RUN!" *Essie raised her voice. Something inside her told her there wasn't much time.*

Fifteen people huddled in the back room of Peebles. There were small children. Everyone was calm and quiet. Essie was not. She was concerned she hadn't gotten everyone back there. Was anyone still in the front of the store? She needed to check.

Essie ran back up to the front quickly. She couldn't believe what she saw.

"I was about five feet from the front door. I looked out to see if I could see anything. . . . I saw it then. It was a cloud of black smoke coming up the hill. And I could hear it. It sounded like a thousand trains. In half of a second I had enough time to get under the front *desk. And by that time, it hit. It sounded like the building was going to explode any minute and you just heard . . . rumbling! Banging! And hitting! . . . It got so dark. Tables flew over where I was. . . . Man, you could hear the glass breaking. All the cologne bottles shattered . . . the sounds . . . I've never heard anything like that in my life.*

"That sound was just so excruciating." Essie shudders to remember it. "I just didn't want to hear that noise."

It passed in a matter of seconds, though. "Miss Essie!" a man hollered to her from the back of the store when things seemed to calm down. "You okay?" People moved through the wrecked store to get out.

"Yes, yes. I'm fine," Essie called to him. Some of the salesgirls were crying. But no one was hurt. As the group made their way outside, they saw clothing from the store strewn about the street. All the surrounding storefront shop windows were shattered. Roofs were torn off. Trees and power lines were down. Buildings were demolished. Despite the scene straight out of Armageddon, everyone made it out of Peebles, and they were all fine. People were hugging each other and calling loved ones.

"Then the sky got clear," Essie recalls incredulously. "It was sunshiny and cool. Up above it looked as though nothing had ever happened."

Despite the quick change in the weather that erased all evidence of nature's wrath, Essie says, "I'll never forget that day—April 24, 2010. That's something that will always stick in my mind."

Essie and the people in her store escaped that tornado safely. But others were not so lucky. According to the National Weather Service, ten people were killed by the severe weather that struck Mississippi that day. The preliminary report described a tornado with a path of destruction over 100 miles long and 1.75 miles wide. It was classified as an EF–4 tornado, with maximum winds of over 170 mph.[2]

ESSIE HENDRIX KNEW WHAT TO DO BECAUSE SHE LIS-tened to the NOAA Weather Radio meteorologist. She described the warnings she heard.

SEVERE WEATHER STATEMENT

NATIONAL WEATHER SERVICE JACKSON MS

1203 PM CDT SAT APR 24 2010

...A TORNADO WARNING REMAINS IN EFFECT UNTIL 1230 PM CDT FOR
CENTRAL YAZOO COUNTY ...

...THIS IS A TORNADO EMERGENCY FOR THE WARNED AREA ...

AT 1203 PM CDT ... NATIONAL WEATHER SERVICE METEOROLOGISTS
AND STORM SPOTTERS WERE TRACKING A LARGE AND EXTREMELY
DANGEROUS WEDGE TORNADO. THIS TORNADO WAS LOCATED 6 MILES
NORTH OF SATARTIA MOVING NORTHEAST AT 60 MPH.

THE TORNADO WILL BE NEAR ...

YAZOO CITY AND LITTLE YAZOO BY 1210 PM CDT ...

MYRLEVILLE BY 1215 PM CDT ...

BENTON AND EDEN BY 1220 PM CDT ...

MIDWAY BY 1225 PM CDT ...

PRECAUTIONARY/PREPAREDNESS ACTIONS ...

A TORNADO WARNING MEANS THAT A TORNADO IS OCCURRING OR
IMMINENT. YOU SHOULD ACTIVATE YOUR TORNADO ACTION PLAN AND
TAKE PROTECTIVE ACTION NOW. SIGNIFICANT DAMAGE HAS OCCURRED
WITH THIS SIGNIFICANT TORNADO!

THIS IS AN EXTREMELY DANGEROUS AND SERIOUS LIFE-THREATENING
SITUATION. THIS STORM IS CAPABLE OF PRODUCING STRONG TO
VIOLENT TORNADOES. IF YOU ARE IN THE PATH OF THIS TORNADO ...
TAKE COVER IMMEDIATELY!

There are specific terms to break down in this statement. A tornado emergency is an exceedingly rare type of tornado warning that is issued by the National Weather Service. It means that there is a severe threat to human life and catastrophic damage from an imminent or ongoing tornado. It is a high-impact call to action, and is reserved for situations where there is clear radar evidence or visual observation of the tornado.[3]

CRITERIA

- A large and catastrophic tornado has been confirmed and will continue. (A radar signature alone is not sufficient.)
- It is going to have a high impact and/or affect a highly vulnerable population
- There is a potential for numerous fatalities[4]

Also mentioned in the above advisory was the term "wedge tornado." Here is what that means, according to NOAA:

WEDGE TORNADO

"Wedge" is an informal phrase for a **tornado** that *looks* wider than the distance from ground to cloud base. There is no scientific meaning to it, but many meteorologists both in the National Weather Service and in the media use this term to describe especially violent and wide tornadoes. Many factors, such as actual tornado size, cloud base height, moisture content of the air, intervening terrain, soil, and dust lofting can regulate a tornado's apparent width. Although many famous wedge tornadoes have also been violent, a tornado's size does not necessarily indicate anything about its strength.

Severe thunderstorms are capable of producing tornadoes at any time. Meteorologists from the National Weather Service are carefully looking for any indication that these thunderstorms are showing signs of developing tornadoes, using Doppler radar and a network of storm spotters that can offer visual reports. When a tornado is indicated, forecasters issue the appropriate advisories to the public. This is something you may hear on local TV stations or on your NOAA Weather Radio. It is important to keep an eye and ear on these sources of information when severe weather is threatening—they will be the first to tell you if a tornado is approaching. (For more on severe thunderstorms, see Chapter 9.)

Important tornado advisory terms to know are: *tornado watches* and *tornado warnings*. Note the distinction between the two:

> **Tornado Watch:** This is issued by the National Weather Service **when weather conditions are favorable for the development of tornadoes in and close to the watch area.** Their size can vary depending on the weather situation. They are usually issued for four to eight hours and well in advance of the actual occurrence of severe weather. During the watch, people should review tornado safety rules and be prepared to move to a place of safety if threatening weather approaches.

> **Tornado Warning:** This is issued when a tornado is indicated by radar or has actually been sighted by spotters; therefore there is immediate danger. **People in the affected area should seek safe shelter immediately.** A tornado warning can be issued without a tornado watch being already in effect. The warning is usually issued for 30 minutes. If a tornado has not hit in that time, it likely has passed.

RADAR

When conditions are right for severe thunderstorms, the National Weather Service monitors the storms on radar. Sometimes tornado warnings are issued based on radar detections, not just on spotters' sightings.

The Brandon, Mississippi, Doppler radar showed the tornadic thunderstorm at 12:14 p.m. (CDT) on April 24, as the storm was moving across Yazoo City. An area of higher reflectivity, resulting from significant tornado debris being lifted into the atmosphere, called a "debris ball," was also evident. Radar also showed intense rotation in the storm using the "Doppler" product, which can detect which way the wind is blowing relative to where the storm is. (NOAA).

RADAR EVIDENCE

A thunderstorm that is producing a tornado, when observed by radar, has certain distinguishing features. Computer algorithms analyze Doppler radar data and display it in ways that make it easier to identify dangerous weather. Forecasters are trained to recognize the precise radar signatures produced automatically by these sophisticated computer applications. Today's weather radars typically provide, on average, 11 minutes of lead time, and can pinpoint locations directly in harm's way with a high degree of accuracy.

—Source: National Severe Storms Laboratory

WHAT ARE TORNADOES?

A tornado, as defined by NOAA, is a violently rotating column of air that extends from the base of the thunderstorm to the ground. It nearly always starts as a funnel cloud that has not yet touched the ground and may be accompanied by a loud roaring noise. On a local scale, it is **the most destructive of all atmospheric phenomena.**

Although tornadoes occur in many parts of the world, they are found most frequently in the United States east of the Rocky Mountains during the spring and summer months. **According to the NOAA, in an average year, 800 tornadoes are reported nationwide, resulting in 80 deaths and over 1,500 injuries.**

Tornadoes are the most violent of all atmospheric storms.

There are two types of tornadoes: those that come from a *supercell thunderstorm,* and those that do not.[5]

Tornadoes that form from a supercell thunderstorm are the most common and often the most dangerous. A supercell is a long-lived (longer than one hour) and highly organized storm feeding off an updraft (a rising current of air) that is tilted and rotating. This rotating updraft—as large as ten miles in diameter and up to 50,000 feet tall—can be present for as long as 20 to 60 minutes before a tornado forms. Most large and violent tornadoes come from supercells.

Non-supercell tornadoes are circulations that form without a rotating updraft. One non-supercell tornado is the *gustnado,* a whirl of dust or debris at or near the ground with no condensation (rain) funnel, which forms along the gust front of a storm. Another non-supercell tornado is a *landspout.* A landspout is a tornado with a narrow, ropelike condensation funnel that forms when the thunderstorm cloud is still growing and there is no rotating updraft—the spinning motion originates near the ground. *Waterspouts* are similar to landspouts, except they occur over water.[6]

A waterspout normally refers to a small, relatively weak rotating column of air over water beneath a towering cumulus cloud. Waterspouts are most common in warm and tropical or subtropical waters. There are two types:

- Tornadic waterspouts generally begin as true tornadoes over land in association with a thunderstorm, and then move out over the water. They can be large and are capable of considerable destruction.
- Fair-weather waterspouts form directly over open water. They climb skyward in association with warm water

temperatures and high humidity close to sea level. They are usually small, relatively brief, and less dangerous. Fair-weather waterspouts are much more common than tornadic waterspouts.

—Source: NOAA's National Severe Storm Laboratory

WHERE DO TORNADOES COME FROM?

Tornadoes come from the energy released in a thunderstorm. As powerful as they are, tornadoes account for only a tiny fraction of the energy in a thunderstorm. What makes them dangerous is that their energy is concentrated in a small area, perhaps only a hundred yards across. Not all tornadoes are the same, of course, and science does not yet completely understand how part of a thunderstorm's energy sometimes gets focused into something as small as a tornado.

—Source: National Severe Storms Laboratory

HOW DO TORNADOES FORM?

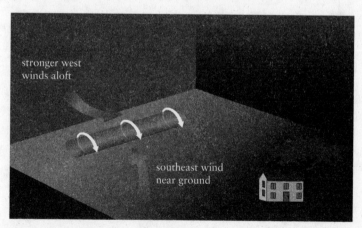

Before thunderstorms develop, a change in wind direction and an increase in wind speed with increasing height create an invisible, horizontal spinning effect in the lower atmosphere.

Rising air within the thunderstorm updraft tilts the rotating air **from horizontal to vertical.**

An area of rotation, two to six miles wide, now extends through much of the storm. Most strong and violent tornadoes form within this area of strong rotation.

TORNADOES CAN OCCUR AT ANY TIME OF THE YEAR

- In the southern states, peak tornado season is March through May.
- In the northern states peak season is summer.
- Tornadoes are most likely to occur between 3 and 9 p.m. but have been known to occur at all hours of the day or night. Tornadoes can last from several seconds to more than an hour.
- The longest-lived tornado in history is unknown, because so many of the long-lived tornadoes reported from the early to mid-1900s and before are believed to be tornado series instead.
- Most tornadoes last less than ten minutes.
- The average tornado moves from southwest to northeast, but tornadoes have been known to move in any direction. The average forward speed is 30 mph but may vary from nearly stationary to 70 mph.

—Source: NOAA Storm Prediction Center

THE ENHANCED FUJITA (EF) SCALE

Dr. Tetsuya Theodore Fujita first introduced the Fujita Scale in February 1971, according to the Storm Prediction Center. Fujita wanted to create something that categorized each tornado by intensity and area. In 2007, the Fujita Scale was updated to the Enhanced Fujita Scale. Roger Edwards, who wrote the Online Tornado FAQ for the Storm Prediction Center, explains why the new EF scale is used today:

The idea is that a "one size fits all" approach just doesn't work in rating tornado damage, and that a tornado scale needs to take into account the typical strengths and weaknesses of different types of construction. This is because the same wind does different things to different kinds of structures. The Enhanced F Scale is a much more precise and robust way to assess tornado damage than the original.

It classifies F0–F5 damage as calibrated by engineers and meteorologists across 28 different types of damage indicators (mainly various kinds of buildings, but also a few other structures, as well as trees).[7]

The Enhanced Fujita Scale		
Rating	Description	Wind speed
EF0	Gale tornado	65–85 mph
EF1	Moderate tornado	86–110 mph
EF2	Significant tornado	111–135 mph
EF3	Severe tornado	136–165 mph
EF4	Devastating tornado	166–200 mph
EF5	Incredible tornado	> 200 mph

—*Courtesy NOAA*

TORNADO SAFETY

Know the signs of a tornado: There is no substitute for staying alert to the sky. Besides an obviously visible tornado, here are some things to look and listen for:

1. Strong, persistent rotation, or spinning in the cloud base.
2. Whirling dust or debris on the ground under a cloud base—sometimes tornadoes have no funnel cloud!
3. Hail or heavy rain followed by either dead calm or a fast, intense wind shift. Many tornadoes are wrapped in heavy rain and can't be seen.
4. Day or night: loud, continuous roar or rumble (some describe the sound similar to a freight train); this doesn't fade in a few seconds like thunder.
5. Night: small, bright, blue-green to white flashes at ground level near a thunderstorm. These mean power lines are being snapped by very strong wind, maybe a tornado.

TORNADO COLORS

Pictures of tornadoes show them looking white, gray, and black or even red or pink. Why is that? It has to do with the viewing angle and the color of the dirt and debris the tornado picks up. The tornadoes tend to look darkest when looking southwest to northwest in the afternoon. In those cases, they are often silhouetted in front of a light source, such as brighter skies west of the thunderstorm. If there is heavy precipitation behind the tornado, it may be dark gray, blue, or even white, depending on where most of the daylight is coming from. Tornadoes wrapped in rain may exhibit varieties of gray, if they are visible at all. Lower parts of tornadoes also can assume the color of the dust and debris they are generating; for example, a tornado passing across dry fields in western or central Oklahoma may take on the hue of the red soil so prevalent there.

—Source: SPC

6. Night: *persistent* lowering from the cloud base, illuminated or silhouetted by lightning—especially if it is on the ground or there is a blue-green-white power flash underneath. (Note: Having easy access to your battery-powered NOAA Weather Radio is especially true overnight. An alarm will come on to alert you while you are sleeping. In the event that a tornado warning is issued, you need to take cover. Many tornadoes strike at night.)

—Source: SPC, NOAA, FEMA, REDCROSS.org

EXCLUSIVE INTERVIEW WITH A TORNADO EXPERT

Don Burgess is considered one of the most renowned tornado scientists in the United States. For over 30 years, Burgess studied severe storms and tornadoes at the National Severe Storms Laboratory (NSSL) and gained real-world experience by storm chasing. At the NSSL, Burgess also served as director and chief of operations at the NEXRAD Radar Operations Center and as chief of National Weather Service radar training. The Oklahoma native knows tornado damage firsthand; he's performed over 100 tornado damage surveys himself.

Below he answers questions on tornado safety exclusively for readers of this book:

1. What should you do to prepare your home if a tornado threatens?
Each home should have a tornado plan to address the following questions: Is the home strong enough to use for shelter? (**Mobile or manufactured homes should be vacated for stronger shelter if time permits.**) Where will home inhabitants take shelter if a tornado warning is issued? If in the home, the location should be known to all and easy to reach. If the location is outside the home, the plan to

get to the location should be known by all, to ensure that everyone can reach the location. Each home should have a tornado kit that goes with them to their shelter.

New, site-built homes can have added protection from tornado winds built into construction for just a few thousand dollars. Older homes can have added wind resistance added, but the costs are higher. In-home and underground tornado shelters can be purchased. If a home has an underground or in-home storm shelter, the location should be made known to local law enforcement, emergency management, and neighbors so that if debris blocks exits from the shelter, rescuers will know where to dig out those in the shelter.

2. What actions should you take if you are at home and there is a tornado warning?
If a tornado warning is issued, home residents need to decide if they will shelter in place or seek stronger shelter. If seeking stronger shelter, is there time to reach that shelter? **You don't want to be caught in the open or in a car—that is a worst-case scenario.** If seeking shelter in place, go to the shelter location quickly and seek additional information from easy-to-obtain radio sources.

3. What are the misconceptions people have about tornadoes, and what are common mistakes people often make?
Some people believe that weak infrastructures such as mobile homes will protect them from tornadic winds. Many people still believe that they should open windows in their homes to help equalize pressure differences as a tornado passes over the house. **We have learned that pressure will equalize without the need to open windows.**

Other people believe that staying in cars or lying in open, shallow depressions will protect them from tornado winds. **Cars are not safe in tornadoes, and those people lying in the open are subjected to the chance of injury from wind-blown debris.**

4. What are some of the most important safety supplies people should have at home?
Each home should have a tornado kit that consists (at a minimum) of a flashlight, first-aid kit, commercial radio, NOAA Weather Radio, and whistle or some other sound-making device that would alert rescuers to victim location. **Home owners should know that cell phones can fail easily if the towers are damaged or networks become overloaded and useless.**

5. If a tornado is truly coming and someone had only minutes to take action, what should they do?
If quick shelter in a home is needed, those seeking shelter should:

- get away from windows and outside walls
- seek the lowest level and central part of the structure
- go to central rooms, such as bathrooms and closets, which can provide good shelter (bathrooms are especially good because of the added reinforcement of walls provided by water pipes)
- shelter under protective cover (stairways, heavy tables, workbenches, etc.)
- try to get under a mattress or be covered by blankets or some other insulating material to help protect against blowing debris, including glass.

Going to a central bathroom, getting down in a bathtub, and covering with a mattress have saved many tornado

victims. I've witnessed the safety of such shelter in
destroyed homes many times.

**6. What should you do if you are driving and you
hear on the radio that your county or parish is under a
tornado warning?**
If you are driving and you become aware of a tornado
warning, you should seek better shelter than in a vehicle
or you should drive to a safe location. If you are in a
metropolitan area where traffic moves slowly and can
be stopped, or if you can't see/don't know where the
tornado is located, you should **drive to the closest well-
built and available structure and seek shelter inside.**
Examples would be buildings like libraries, schools,
churches, office buildings, or hospitals. If you are in a
rural area with no impediments to higher speed travel
(i.e., a highway network) and you can see the tornado in
the distance or you know where the tornado is located
and which way it is moving, you might be able to
move to a safe location by driving at right angles to the
expected tornado path. **A vehicle is not safe shelter.** As a
last resort, if all other options are unavailable, you can
leave the vehicle and lie down in a ditch with steep sides
or, better yet, hide in a culvert or large drainage pipe.
**If shelter is sought in a low, ditchlike place, remember
that thunderstorms produce heavy rain and low places/
ditches may flood. Stay away from any potential high
water location.**

**7. People often gather (especially motorcyclists) under
highway girders and overpasses during storms. Is that a
good idea or bad?**

Gathering under highway overpasses is a bad idea. First, tornado winds funnel through such locations and many times blow people away from girder areas or pelt them with windblown debris. Second, vehicles stopping under overpasses create traffic jams that partially or completely block highways, creating traffic hazards and putting large numbers of drivers at risk.

8. What should you do if you are in a busy public area and a tornado is coming?

In a public area, seek shelter in a well-built building. If in an arena or stadium, seek shelter in an underneath location such as a bathroom. If in an open area like a park, a public bathroom may provide shelter if it's well built. As a last resort in an open area, lie in a ditch or drainage pipe. If your location is open to the sky, **cover your head as debris will fall on you.**

9. Can you describe in your experience, from what you've seen in terms of specific tornadoes destruction, why taking safety measures is so important?

The most severe tornadoes (EF4 and EF5 on the Enhanced Fujita Scale) can destroy even well-built houses. The only really safe place in a violent tornado is an aboveground or underground safe room or shelter or a well-built larger building. **I have seen well-built homes, apartment buildings, strip malls, small commercial or office buildings, and manufacturing facilities completely wiped away, leaving only the concrete foundation.** During a survey of the terrible May 3, 1999, Oklahoma City–area tornado, I saw all of those building types destroyed as well as whole mobile parks obliterated and thousands of cars thrown large distances. **Fortunately,**

only a few tornadoes a year in the United States get that
strong. Most tornadoes are weaker, and my experience
is that well-built homes provide safe shelter. But a well-
built structure is a must; I have seen over and over the
tornado wind casualties caused by trying to shelter in
barns, mobile homes, vehicles, and other unsafe places.

**10. After a tornado warning is over, what immediate
actions should people take?**
**After a tornado passes, rescuing victims is the most
important activity.** If you are in the immediate damage
area, you might be able to assist in search and rescue or
provide aid to the injured.

- Pay attention to local law enforcement and emergency
  management. If they indicate you can help, stay and do so, but
  if they indicate your presence is adding to confusion, leave as
  directed.
- Never go into areas that smell of leaking natural gas or other
  harmful agents. Be very cautious around downed electrical lines
  that might still be electrified.
- If you are away from damage areas, never drive to the damage
  area to see how much damage might have been done.
- Don't gather in groups just to see what's happening. **"Gawkers"
  get in the way of emergency vehicles and search and rescue.
  They make a bad situation worse.**

MORE TORNADO SAFETY TIPS
(NOAA, FEMA, The Red Cross)

Protect Your Home from Wind Damage:

1. Flying debris is the greatest danger in tornadoes. Make
   sure there are protective coverings near your shelter

space, such as mattresses, sleeping bags, or thick
blankets.

2. Move or secure lawn furniture, trash cans, hanging
plants, or anything else that can be picked up by the wind
and cause injuries

3. Remove diseased and damaged limbs from trees.

4. Move or secure lawn furniture, trash cans, hanging
plants, or anything else that can be picked up by the wind
and cause injuries.

WHAT TO DO IF A TORNADO THREATENS IN VARIOUS SITUATIONS AND LOCATIONS

If you have a basement: Get in the basement and
under some kind of sturdy protection (heavy table or
workbench), or cover yourself with a mattress or sleeping
bag. Know where very heavy furniture, such as a piano,
refrigerator, waterbed, etc., rests on the floor above
you, and avoid being under them. They may fall down
through a weakened floor and crush you.

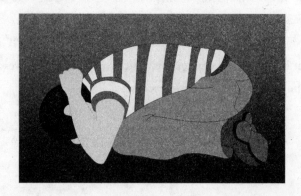

**If you're in a house with no basement, a dorm room, or
an apartment:** Avoid windows. Go to the lowest floor,
a small center room (like a bathroom or closet), under

a stairwell, or in an interior hallway with no windows. Crouch as low as possible to the floor, facing down; cover your head with your hands. A bathtub offers a shell of partial protection. Even in an interior room, you should cover yourself with some sort of thick padding (mattress, blankets, etc.), to protect against flying debris.

In an office building or hospital: Go directly to an enclosed, windowless area in the center of the building, *away from glass,* and on the lowest floor possible. Then crouch down and cover your head. Interior stairwells are usually good places to take shelter, and if not crowded, allow you to get to a lower level quickly. Stay off the elevators; you could be trapped in them if the power is lost.

In a mobile home: Even if your home is tied down, you are probably safer outside, even if the only alternative is to seek shelter out in the open. Most tornadoes can destroy even tied-down mobile homes; and it is best not to play the low odds that yours will make it. If your community has a tornado shelter, go there fast. If there is a sturdy permanent building within easy running distance, seek shelter there. Otherwise, lie flat on low ground away from your home, protecting your head. If possible, use open ground away from trees and cars, which can be blown onto you.

At school: Follow the drill. Go to the interior hall or room in an orderly way as you are told. Crouch low, head down, and protect the back of your head with your arms. Stay away from windows and large open rooms like gyms and auditoriums.

In a car or truck: Vehicles are extremely dangerous in a tornado. If the tornado is visible, is far away, and the traffic is light, you may be able to drive out of its path by moving at right angles to the tornado. Otherwise, park the car as quickly and safely as possible—out of the traffic lanes. Abandon the vehicle and seek shelter in a sturdy building. Avoid seeking shelter under bridges, which can create deadly traffic hazards while offering little protection against flying debris.

In the open outdoors: If possible, seek shelter in a sturdy building. If not, lie flat and facedown on low ground, protecting the back of your head with your arms. Get as far away from trees and cars as you can; they may be blown onto you in a tornado.

In a shopping mall or large store: Do not panic. Move as quickly as possible to an interior bathroom, storage room, or other small, enclosed area, away from windows.

In a church or theater: Do not panic. If possible, move quickly but orderly to an interior bathroom or hallway, away from windows. Crouch facedown and protect your head with your arms. If there is no time to do that, get under the seats or pews, protecting your head with your arms or hands.

BUILDING A SAFE ROOM IN YOUR HOME
(FEMA)

Extreme windstorms in many parts of the country pose a serious threat to buildings and their occupants. Your residence may be built "to code," but that does not mean it can withstand winds from extreme events such as tornadoes and major hurricanes. The purpose of a safe room or a wind shelter is to provide a space where you and your family can seek refuge that provides a high level of protection. You can build a safe room in one of several places in your home:

- your basement
- atop a concrete slab-on-grade foundation or garage floor
- an interior room on the first floor

Safe rooms built below ground level provide the greatest protection, but a safe room built in a first-floor interior room also can provide the necessary protection. Below-ground safe rooms must be designed to avoid accumulating water during the heavy rains that often accompany severe windstorms.

To protect its occupants, a safe room must be built to withstand high winds and flying debris, even if the rest of the residence is severely damaged or destroyed. Consider the following when building a safe room:

- The safe room must be adequately anchored to resist overturning and uplift.
- The walls, ceiling, and door of the shelter must withstand wind pressure and resist penetration by windborne objects and falling debris.
- The connections between all parts of the safe room must be strong enough to resist the wind.
- Sections of either interior or exterior residence walls that are used as walls of the safe room must be separated from the structure of the residence so that damage to the residence will not cause damage to the safe room.

More information on how to build a safe room can be found at: http://www.fema.gov/hazard/tornado/to_saferoom.shtm and www.flash.org (The Federal Alliance for Safe Homes).

TORNADO SAFETY FOR PETS

Dr. Scott Mason is a veterinarian and the disaster preparedness coordinator for the Oklahoma Veterinary Medical Association.

"Pet owners should make preparations with their pets before severe weather season," he advises. "Preparedness is everything. Have your emergency kit prepared and readily accessible."[8]

(Tips on how to prepare a Family Disaster Plan for People and Pets and how to create a Pet Emergency Supply Kit can be found in Chapters 13 and 15, respectively.)

> Owners need to understand that their safety needs come first. It won't do them any good to not seek shelter and be injured while trying to protect their pet. If a situation arises, such as being unable to gather a scared cat, they should always protect themselves first and then after the event they will be able to help their pet.
>
> —Dr. Scott Mason, DVM

Here are his tornado safety tips for pet owners:

- Evacuate with your pets—especially from a mobile home.
- If you live in an apartment or regular home, you should have a selected room or practice evacuation as well. For pets, this should include ready access to your preparedness kit, including appropriate leashes and/or cages. Efforts should begin as tornado watches are issued. **This is especially important for animals that may be difficult to catch, such as nervous cats.**
- Practicing this procedure with your pets will allow them to become used to the action and be less stressed by the situation.
- Positive reinforcement of the behavior should always be given. The pets should learn that this is a good thing and they will be rewarded for it.

- **Keep in mind that tornadoes are stressful to animals. They are scared, are nervous, and exhibit behaviors that could be considered abnormal for them (such as biting and scratching people, even their owners). Care should always be exercised in approaching these animals.**
- Most dogs, if not with their owners, will be fearful and defensive. Some will remain near their familiar territory while others will roam and seek to explore or find a secure location to hide.
- Cats will also be fearful and defensive. They tend to escape and find a secure location/hiding spot. They will usually hide during the day and return to familiar territory, usually at night, to investigate and seek food and water.
- Do not reinforce nervousness and anxiety. **Pets will be aware of their owners' anxiety and will exhibit stress themselves because of this.** Coddling and excessive comforting reinforce to the pets that they have good reason to be fearful. The sheltering should be calm, quiet, routine, and matter-of-fact.
- Reinforce only calm, good behavior. Owners should reinforce good behavior with rewards such as favorite toys and treats.
- Nervous cats and dogs in cages can sometimes feel more secure if the cage is covered with a towel or blanket (obviously still allowing for air circulation).

AFTER THE DISASTER

- Monitor your pets closely and keep them leashed. Familiar scents and landmarks may be altered and can cause confusion or abnormal behavior.
- Examine your animals closely, and contact your veterinarian immediately if you observe injuries or signs of illness.
- Survey the area inside and outside your home to identify potential dangers (downed power lines, fallen trees, debris).

- Release pets indoors only. They could encounter dangerous hazards if allowed to go outside unsupervised.
- Reintroduce food in small servings and gradually work up to full portions if your pets have been without food for a prolonged period of time.
- Allow uninterrupted rest/sleep for all animals to recover from the trauma and stress.

CONCLUSION

Tornadoes are violently rotating columns of air, and one of the most destructive and life-threatening weather phenomena. The most dangerous tornadoes develop from supercell thunderstorms. Tornadoes can also form over the ocean, in which case they are called waterspouts. While tornado watches may give you some time to prepare, as they may be issued hours ahead of the tornado's potential to touch down, a tornado warning means you may have only minutes to take cover. The safest place to be is underground, in a basement. If your home doesn't have a basement, seek the most interior room of your home, like a bathroom or closet. Bring pillows, covers, and mattresses (if possible) to protect yourself from flying or falling debris, as many injuries occur when furniture and other items become projectiles. Tornadoes are scary not only for people, but for animals as well. These dangerous storms may cause family pets to exhibit abnormal behavior. Extra caution and care are advised when dealing with animals, both before and after the storm.

3
FLASH FLOODS

So ready to go camping this weekend. Kaden is going to love it![1]

wrote 23-year-old Leslie Jez in the status box of her Facebook page
early on Monday, June 11, 2010. She later added:

*Not looking forward to that cold water, but sounds like I
might change my mind after seeing how hot it's supposed
to be.*[2]

Leslie was a self-described cowgirl. Growing up in rural Arkansas,
Leslie helped raise quarter horses and round up cattle on her grandparents' ranch. Her passion for country living was shared by her husband,
Adam, and their three-year-old son, Kaden. The toddler already had
his own baby cowboy boots.[3]

The plan was for the family, which included Leslie, Adam, and
Kaden, plus Leslie's mother, brother, and sister, to spend the weekend camping. They were headed for the Albert Pike Recreation Area
in Cappo Gap, Arkansas, located in the Ouachita National Forest,
which sits in a valley between two heavily forested hills near the Little
Missouri River.[4]

The Jez family, along with about two to three hundred other people, camped outside in tents, cabins, and RVs. At around 11 p.m. that
evening, light rain began to fall. As most campers slept into the early
hours of Friday morning, the downpours became torrential. Soon the
storm caused the Little Missouri River to rise and gush water down-
ward onto the campsite.

This happened at an incredibly rapid pace. By 2 a.m. the campground was under four feet of water; three hours later it was under 23
feet of water.[5] Trees were uprooted. Tents washed away, and RVs rolled
onto their sides and floated like boats. Leslie's sister Haley had to cling
to a tree to save her life.[6]

Leslie and little Kaden were not so lucky; both were killed when their RV was swept away by the floodwaters. Leslie's husband, Adam, and her brother Randall survived. Leslie's mother, 47-year-old Sherry, did not.

Twenty people drowned in the flash flood at Albert Pike in Arkansas that June day not long ago. For years people had camped in this very site, but the water had never risen to such extremes. The disaster was compounded because it happened in the middle of the night while most people were asleep.

FLASH FLOODS ARE A TOP WEATHER-RELATED KILLER IN the United States. While campsites, because of their remote location away from sturdy buildings, are a dangerous place to be during a flash flood, the most deadly spot to be caught is in your car. **More than half of all flood deaths occur in vehicles.**

Here is what to do when you are in a flash flood in a variety of places:

IN YOUR CAR

The first point is to avoid the situation, so you're not in your car during a flash flood. Be aware of the weather conditions before you go. Most flash flooding is caused by slow-moving thunderstorms, heavy rain from hurricanes or tropical storms, snowmelt, or ice jams in rivers.

Listen specifically to your NOAA weather radio or local TV weather broadcasts for advisories like:

> **Flash Flood Watch:** Flooding is possible within the watch area.
>
> **Flash Flood Warning:** Flooding has been reported or is imminent.
>
> **Urban and Small Stream Advisory:** Flooding of small streams, streets, and low-lying areas, such as railroad underpasses and storm drains, is occurring.

NOAA advises that if you see water covering a road: "Turn Around, Don't Drown." This seems obvious, but many people try to drive across a flooded road, assuming it's not too deep, only to be proven dead wrong. Water depth is particularly tough to judge at night. Do not attempt to cross a flooded road, even if the car in front of you made it through.

Even very shallow water can move a car. People often underestimate the force of moving water.

- As little as six inches of water can move a car. It reaches the bottom of most passenger vehicles, causing them to lose control or stall.
- As little as a foot of water is enough to float a vehicle.
- Two feet of rushing water can carry away most larger vehicles, including SUVs and pickups.

WHAT TO DO WHEN YOUR CAR STALLS ON A FLOODED ROAD

When you're in the ocean and a wave breaks on you, it can knock you off your feet. Then the water gets calm, so you can stand back up again. The force of moving floodwater is not like that; it's constant. There is no lull where you can get back up and recover.

—Ricky Slack, director of Smith County Emergency Services and
the captain of their local SWIFT water rescue team

Captain Slack has rescued dozens of people caught in flash floods. In an exlusive interview for this book, he says people typically underestimate the force of moving floodwater.

His advice, along with NOAA's, is included in many of these tips on what to do in the dire situation of being caught in your car in the rushing water of a flash flood:

- Power the windows down quickly before the car's electrical system shorts out. Your first means of escape is the door, but the window may become the next option.

- Abandon the vehicle to reach higher ground. The water may rise further, and the most important thing is to get you, not the car, to safety.
- Don't worry about gathering belongings inside the car; they can be replaced.
- If you cannot open the door or roll down the window to get out, use a narrow, but sharp, object to break the window so you can climb out. A tool called a "life hammer" or a "window punch" can do this. (A complete list of safety items to have in your car is in the Emergency Supply List for People in chapter 14.) The SWIFT water rescue team uses a similar tool in their work.
- When exiting the vehicle, make sure you have a clear path to higher terrain. If that isn't possible, climb on top of the car and hold on for support.
- If you have a cell phone with you with a GPS feature, use that to notify authorities where you are.
- In the worst-case scenario, where the car is filling with water and you are trapped inside, the pressure of the water inside can give you the leverage to force the doors open so you can swim to the surface.

WHAT TO DO WHEN YOU ARE CAUGHT
ON FOOT IN A FLASH FLOOD

- Be aware of any flash flood watches, warnings, or advisories before you venture out.
- If you are walking on a street that starts to flood, turn around and go in the opposite direction. Do not attempt to walk through water, as the depth is misleading. It takes only six inches of water to knock you off your feet.
- If you must walk through water, walk where the water is still and not moving. If you can find one, use a stick to determine the depth of the water in front of you.

- In the worst-case scenario, where you are on foot and swept up by floodwaters, Captain Slack advises that you get into a "recovery position": "butt down, on your back so you can see your toes and float."
- While floating on your back, point your toes downstream; this will prevent your head from striking any object. Try to grab anything you can to hold you in place, like a large rock or a tree.
- If you feel that you can swim toward an elevated area, Captain Slack advises that you swim upstream diagonally. That's the best way to fight the current of moving floodwater.
- Try to pull yourself or most of your body out of the water as quickly as possible. This is important to avoid hypothermia.

WHAT TO DO WHEN YOU ARE AT HOME
AND THERE IS A FLASH FLOOD

PREPARE

- Know your home's flood risk, particularly its location and elevation above flood stage.
- Have a NOAA weather radio, batteries, and all supplies, including nonperishable food and bottled water, ready. (A complete list of emergency supplies for people can be found in chapter 14, and for pets in chapter 15.)
- Sandbags are supplied for free by many counties to protect your home. Contact your local government emergency management to find out how to get them.
- Turn off all utilities at the main power switch, and close the main gas valve if advised to do so.
- Install check valves in building sewer traps to prevent floodwater from backing up into the drains of your home.
- Remove or seal all hazardous products from your home.

- If possible, to protect your furniture and if there is time, roll up rugs and curtains from the floor and move furniture away from walls.
- Secure all backyard furniture, toys, etc. These items can become dangerous to others if they are caught in floodwaters.
- Unplug all electrical appliances. Bring as many of them as you can, along with all valuables, upstairs or to the highest level of your home.

DURING

- The top floor is where you should go as well during the flood threat. Don't wait it out in the basement or bottom level.
- If you've come into contact with floodwaters, wash your hands with soap and disinfected water.
- If you are ordered to evacuate, comply. It is for your own health and safety. Floodwater can become toxic and spread disease.

WHAT TO DO WHEN YOU ARE CAMPING AND THERE IS A FLASH FLOOD

U.S. park ranger Craig Thexton is the park search-and-rescue co-ordinator of the Zion National Park in Springdale, Utah. Part of his job is to advise park visitors of potential dangers like flash floods.

> An educated visitor can make sound judgments and keep themselves out of harm's way. Some do not. The park has dealt with people who have been stranded by floods, been seriously injured when swept away by a flood, and even those who have been killed in flood events. Be prepared, pay attention to your surroundings, and use good judgment.
>
> —Craig Thexton, U.S. park ranger[7]

His insights and experience are included in the tips below.

- Check the weather before your trip. This may sound obvious, but it's often forgotten when people are in "vacation mode" headed to escape for a few days.
- Be aware of the extended weather forecast, as well as what is happening in your vicinity. Flash floods can occur miles beyond the rainfall that generated it. In 1997, 11 hikers were killed in the Lower Antelope Canyon of northern Arizona when a flash flood filled the narrow canyon with water that reached a depth of 50 feet. Distant rainfall, thunder, or lightning should all be taken seriously.
- Take a NOAA weather radio and extra batteries with you to be alert to any incoming flood advisories. (A complete list of emergency supplies for camping can be found in the Family Disaster Plan chapters.)
- Know the area. Get a map so that you know the topography up- and downstream. Make sure you are familiar with how to get to higher terrain, if need be.

In terms of setting up camp, here are some tips to minimize flood risk:

- Camp at least a quarter mile away from any water source.
- Look for evidence of past flooding: large logs littering the creek/ riverbed, high-water marks on banks, wounded trees next to the water, or a slow-moving bend in the river next to a low, flat area where water can collect.
- Don't rely on your cell phone. (In the case of the Albert Pike campground disaster in Arkansas, many campers were unable to get cell phone reception.)

Ranger Thexton advises that you always hike with a partner; that way one or more of you can go for help. "They [hiking partners] are the most reliable way to get accurate information to rescuers," he says.

Personal locator beacons have become popular in recent years. Ranger Thexton says, however, that they are not always reliable in very narrow canyons. He also warns, "These devices are meant to be used only where there is a serious threat to someone's life. Inappropriate use is common."

- If you find yourself hiking or camping in a flash flood, do not try to outrun it; always seek higher terrain.
- Stay on higher ground until conditions improve. Water levels typically recede within 24 hours.

HOW TO KEEP YOUR PETS SAFE IN A FLASH FLOOD

Dr. Susan Nelson, a veterinarian and clinical assistant professor at Kansas State University's Pet Health Center, says in her online podcast on pet safety in natural disasters that it's a good idea to be ready well in advance of an emergency for all your pet's needs.[8]

She advises to have ready to go:

- All your pets' medicines with dosing instructions, medical documents sealed in a plastic bag.
- Three days of prepackaged food, so you don't have to lug a large food bag.
- (A complete list of other pet emergency supplies is found in the Disaster Plan in chapter 13).
- IDs on pets, both external (like a collar) and internal (like a microchip)

- In case you ever have to evacuate, have a list of pet-friendly hotels within a 40- to 60-mile radius, potential kennels, or other boarding facilities.
- Have a pet carrier ready for smaller animals. You should have one carrier for each pet.
- Dr. Nelson advises you to line the carriers with newspaper, have two small dishes inside (one for food and one for water), and include a litter box if appropriate.
- For flood protection, carriers should be constructed of wire mesh or plastic. Cardboard will disintegrate in water.
- **Keep in mind that animals react differently under stress. Even the most trustworthy pets may panic, hide, try to escape, or even bite or scratch.** If there is a small toy that your pet likes, take it with you when you leave. This will help the animal feel more comfortable in new surroundings.
- In your sealed pet documents, also include a picture of you and your pet. In case the animal is lost, you will have something on hand to show people. This also helps prove pet ownership.

CONCLUSION

Flash floods are a top weather-related killer. Whether flash floods are the result of rapid snow melt into a stream or from heavy rains (including rains from a tropical cyclone), too much water coming in too quickly can prove to be deadly. Since more than half of all deaths from flash floods occur in vehicles, it's especially important to be mindful of a flood threat while driving. Be aware of any flood advisories for your area and never attempt to drive across a water-covered road. At home, there are steps to take to prepare your home for a flood. Some of these include using sandbags, rolling up rugs and curtains from the floor, and unplugging electrical appliances and bringing as many as possible to the highest level of your home.

Before you set up camp at a campground, look for evidence of prior flooding. Know how to get to higher ground in an emergency. Have your pet carrier ready for your animal in case you have to evacuate. Remember, a cardboard carrier will disintegrate in water, so use one that's made of wire or plastic.

4
EXTREME HEAT

It sounded like he was drunk. Eighteen-year-old Tim Smith[1] was slurring his words as he called out to his brother-in-law Rick Jarvis from inside the hot cramped attic to the outside ground below.[2]

"Which way? Moving left . . ." Tim's voice trailed off to what sounded like gibberish.

"Tim, move to the right!" the 31-year-old Rick yelled up to the younger man.

The two were installing electronic hurricane shutters on a one-story home on Marco Island in Florida. It was the first week of August, at 2:30 in the afternoon. The temperature was 98 degrees, and the humidity was at 100 percent.

Rick had left his telecommunications job several years ago to get into the lucrative business of hurricane shutter installation. He was making over one hundred thousand dollars a year now. The money impressed his young brother-in-law. Just out of high school, Tim was eager to join Rick in his business.

The work wasn't easy. Days started at 6 a.m., and even though the goal was to get the job done by 2 p.m. in order to avoid the hottest hours of the day, it didn't always work out that way.

That particular morning, Rich and Tim were very productive. They unloaded the 80- to 100-pound giant rolls of aluminum from their truck and placed them along with other necessary parts in front of all the home's 16 windows. As they attached the heavy metal to the home's exterior, concrete dust blew back all over their faces and arms. They were sweating hard, so it stuck to them like a powdery paste.

The last phase of installation was to do the wiring. The shutters were designed to open and close electronically. That required someone to crawl up into the attic to attach the wires. The attic was a small space, with only a four-foot clearance, so whoever did that part would have to get on his hands and knees to move around. He would also have to scale the rafters carefully, knowing that if he fell, the pads of insulation that surrounded each rafter would not hold his weight. One wrong move and he could fall through the ceiling of the house.

It was almost 3 p.m. The aluminum shutters were up on every window. All that was left was to wire them up. Rick thought, Well, we might as well get the whole job done today.

Tim immediately volunteered. "Hey, Rick, let me do it! I'm smaller anyway."

Tim stood at five foot seven, and Rick was six foot one. Rick thought about it, and since Tim had done the wiring part on homes they'd teamed up on before, it seemed like a good idea.

Tim took the small stairs and climbed into the attic. Rick began to direct him from outside the house as to where the windows were located. It was only ten minutes later when Rick realized something was terribly wrong.

"Tim was slurring his words. He didn't make any sense," Rick recalls. "I decided I needed to go up there myself and get him out."

Rick took the same small stairs and was shocked to find his young protégé lying on his stomach, drenched in sweat.

"What . . . what are you doing up here?" Tim asked. He seemed confused.

"Time to go, buddy. I think the heat is getting to you."

Once inside the attic, Rick felt the temp was at least 120 degrees.

"Nah, I got it!" Tim protested. "I'm fine. I can do . . ." He didn't finish the sentence.

Rick was close enough now to grab Tim's extended leg.

"Tim, let's go." Rick's voice was firm.

Tim resisted more. "Not done yet!" he said.

Rick grabbed Tim's ankles and literally dragged him out. "I felt like I was rescuing someone who was drowning and they were fighting me," Rick remembers.

Eventually he got Tim down the stairs and sat him under a tree in the shade. As Tim guzzled cold water, he seemed to gain focus. He began to realize what had just happened. He had likely been a victim of heat stroke.

"I think you saved my life, man," Tim told Rick. "Thank you."

HOT AND DEADLY

Of all the natural hazards in the United States, heat is the number one non-severe weather related killer.

> " Unlike the roar of an approaching tornado, heat waves kill with silence. In an average year, about 175 Americans succumb to the effects of summer heat.
>
> —NOAA Office of Climate, Water, and Weather Services

HISTORICAL HEAT WAVE DEATHS
(NOAA)

- According to the National Climatic Data Center, in 1980 in the central and eastern United States, an estimated 10,000 deaths occurred during the droughts and heat waves between June and September (including those heat-stress related), followed by 1995 in the Midwest. The cities of Chicago and Milwaukee experienced hundreds of heat-related deaths.
- In August 2003, a record heat wave in Europe claimed an estimated tens of thousands of lives. Parts of France and England saw record high temperatures.
- Traditionally cool regions: The summer of 2010 saw record highs in Moscow and parts of western Russia, where the long-term daily average temperatures for July range from 65–67°F. That year, daily average July temperatures, which includes nighttime temperatures, soared to 87°F. The exceptional heat over such a long duration, combined with poor air quality from wildfires, increased the number of deaths.

WHAT IS EXTREME HEAT?

" Conditions of extreme heat are defined as summertime temperatures that are substantially hotter or more humid than average for
location at that time of year. Humid or muggy conditions, which add
to the discomfort of high temperatures, occur when a "dome" of high
atmospheric pressure traps hazy, damp air near the ground. Droughts
occur when a long period passes without substantial rainfall. A heat
wave combined with a drought is a very dangerous situation.

—The Centers for Disease Control

NOAA'S WATCH, WARNING, AND ADVISORY PRODUCTS FOR EXTREME HEAT

Each **Weather Forecast Office (WFO)** can issue the following
heat-related products as conditions warrant:

Excessive Heat Outlook: Issued when the potential exists
for an excessive heat event in the next three to seven
days. An outlook provides information to those who
need considerable lead time to prepare for the event, such
as public utilities, emergency management, and public
health officials.

Excessive Heat Watch: Issued when conditions are
favorable for an excessive heat event in the next 12 to
48 hours. A watch is used when the risk of a heat wave
has increased, but its occurrence and timing are still
uncertain.

Excessive Heat Warning/Advisory: Issued when an
excessive heat event is expected in the next 36 hours.
**These products are issued when an excessive heat event
is occurring, is imminent, or has a very high probability**

of occurring. The warning is used for conditions posing a threat to life or property. An advisory is for less serious conditions that cause significant discomfort or inconvenience and, if caution is not taken, could lead to a threat to life and/or property.

HOW FORECASTERS DECIDE WHETHER TO ISSUE EXCESSIVE HEAT PRODUCTS

NOAA's heat alert procedures are based mainly on Heat Index Values. The Heat Index, sometimes referred to as the apparent temperature and given in degrees Fahrenheit, is a measure of **how hot it really feels** when relative humidity is factored with the actual air temperature.

The National Weather Service will initiate alert procedures when the heat index is expected to exceed 105–110°F (depending on local climate) for at least two consecutive days.

IMPORTANT: Since heat index values were devised for shady, light wind conditions, **exposure to full sunshine can increase heat index values by up to 15°F.** Also, **strong winds,** particularly with very hot, dry air, can be extremely hazardous.

THE HAZARDS OF EXCESSIVE HEAT

Heat disorders generally have to do with a reduction or collapse of the body's ability to shed heat. This means it's too hot and humid outside for your body to sweat and cool off. When the body can't cool itself efficiently, or when you lose too much fluid or salt through dehydration or sweating, your body temperature rises and heat-related illness may develop. Heat disorders often result from someone being in the heat too long or when someone has exercised too much for his or her age and physical condition.

HIGH RISK FOR HEAT DANGER
(CDC)

Although anyone at any time can suffer from heat-related illness, some people are at greater risk than others.

- Infants and young children are sensitive to the effects of high temperatures and rely on others to regulate their environments and provide adequate liquids.
- People 65 years of age or older may not compensate for heat stress efficiently and are less likely to sense and respond to a change in temperature.
- People who are overweight may be prone to heat sickness because of their tendency to retain more body heat.
- People who overexert during work or exercise may become dehydrated and susceptible to heat sickness.
- People who are physically ill, especially with heart disease or high blood pressure, or who take certain medications, such as for depression, insomnia, or poor circulation, may be affected by extreme heat.

HEAT SAFETY TIPS

WHAT TO DO BEFORE THE HEAT WAVE

NOAA advises that you make advanced preparations for heat by taking the following precautions:

- Visit your physician for a checkup; know if you have a health condition that may be exacerbated by hot weather so you can take extra precaution.
- Have your air conditioner serviced and/or obtain window fans to cool your house.

- Scope out cool indoor places to spend time on hot summer days, such as a local library, aquarium, or museum.

SAFETY TIPS DURING A HEAT WAVE

- **Slow down.** Reschedule strenuous activities until the coolest time of the day. Children, seniors, and anyone with health problems should stay in the coolest available place, not necessarily indoors.
- **Dress for summer.** Wear lightweight, light-colored clothing to reflect heat and sunlight.
- **Put less fuel on your inner fires.** Foods like meat and other proteins that increase metabolic heat production also increase water loss.
- **Drink plenty of water or other nonalcoholic or decaffeinated fluids.** Your body needs water to keep cool. Drink plenty of fluids, even if you don't feel thirsty. People who have epilepsy or heart, kidney, or liver disease, are on fluid restrictive diets, or have a problem with fluid retention should consult a physician before increasing their consumption of fluids. **Do not drink alcoholic beverages, and limit caffeinated beverages.**
- **During a period of excessive heat, spend more time in air-conditioned places.** Air conditioning in homes and other buildings markedly reduces danger from the heat. If you cannot afford an air conditioner, go to a library, store, or other location with air conditioning for part of the day.
- **Don't get too much sun.** Sunburn reduces your body's ability to dissipate heat.

Here are additional hot weather tips from the CDC:

To protect your health when temperatures are extremely high, re-member to keep cool and use common sense. The following tips are important:

Wear Appropriate Clothing and Sunscreen

Wear as little clothing as possible when you are at home. Choose lightweight, light-colored, loose-fitting clothing. Sunburn affects your body's ability to cool itself and causes a loss of body fluids. It also causes pain and damages the skin. If you must go outdoors, protect yourself from the sun by wearing a wide-brimmed hat (which also keeps you cooler) along with sunglasses, and by putting on sunscreen of SPF 15 or higher (the most effective products say "broad spectrum" or "UVA/UVB protection" on their labels) 30 minutes before going out. Continue to reapply it according to the package directions.

Schedule Outdoor Activities Carefully

If you must be outdoors, try to limit your outdoor activity to morning and evening hours. Try to rest often in shady areas so that your body's thermostat will have a chance to recover.

Take Extra Caution If You Exercise or Work Outdoors

If you are not accustomed to working or exercising in a hot environment, start slowly and pick up the pace gradually. If gasping for breath, STOP all activity. Get into a cool area, or at least into the shade, and rest, especially if you become lightheaded, confused, weak, or faint.

Stay Cool Indoors

Stay indoors and, if at all possible, stay in an air-conditioned place. If your home does not have air conditioning, go to the shopping mall or public library—even a few hours spent in air conditioning can help your body stay cooler when you go back into the heat. Call your local health department to see if there are any heat-relief shelters in your area. Electric fans may provide comfort, but when the temperature is in the high 90s, fans will not prevent heat-related illness. Taking a cool shower or bath or moving to an air-conditioned place is a much better way to cool off. Use your stove and oven less, to maintain a cooler temperature in your home.

Other conditions related to risk include age, obesity, fever, dehydration, heart disease, mental illness, poor circulation, sunburn, and prescription drug and alcohol use.

Because heat-related deaths are preventable, people need to be aware of who is at greatest risk and what actions can be taken to prevent a heat-related illness or death. The elderly, the very young, and people with mental illness and chronic diseases are at highest risk. However, even young and healthy individuals can succumb to heat if they participate in strenuous physical activities during hot weather.

Use a Buddy System

When working in the heat, monitor the condition of your coworkers and have someone do the same for you. Heat-induced illness can cause a person to become confused or lose consciousness.

> If you are 65 years of age or older, you should have a friend or relative call to check on you twice a day during a heat wave. If you know someone in this age group, check on them at least twice a day.

When Traveling to a Hotter Climate

Be aware that any sudden change in temperature, will be stressful to your body. You will have a greater tolerance for heat if you limit your physical activity until you become accustomed to the heat. If you travel to a hotter climate, allow several days to become acclimated before attempting any vigorous exercise, and work up to it gradually.

CHILD SAFETY TIPS
(NOAA)

- **Make sure your child's safety seat and safety belt buckles aren't too hot** before securing your child in a safety restraint system, especially when your car has been parked in the heat.
- **Never leave your child unattended** in a vehicle, even with the windows down.
- **Teach children not to play** in, on, or around cars.
- **Always lock car doors and trunks**—even at home—and keep keys out of children's reach.
- **Always make sure all children have left the car** when you reach your destination. Don't ever leave sleeping infants in the car!

CHILDREN, ADULTS, AND PETS ENCLOSED IN PARKED VEHICLES ARE AT GREAT RISK
(NOAA)

Each year children die from hyperthermia as a result of being left in parked vehicles. Hyperthermia is an acute condition that occurs when the body absorbs more heat than it can dissipate. Hyperthermia can occur even on a mild day. Studies have shown that the temperature inside a parked vehicle can rapidly rise to a dangerous level for children, adults, and pets. **Leaving the windows slightly open *does not* significantly decrease the heating rate.** The effects can be more severe on children because their bodies warm at a faster rate than adults'.

HOW FAST CAN THE SUN HEAT A CAR?
(NOAA)

The atmosphere and the windows of a car are relatively "transparent" to the sun's shortwave radiation and are warmed only a small amount. This shortwave energy, however, does heat objects it strikes. **For example, a dark dashboard or seat can easily reach temperatures from 180 to more than 200°F.** These objects (dashboard, steering wheel, child seat) heat the adjacent air by conduction and convection and also give off longwave radiation, which is very efficient at warming the air trapped inside a vehicle.

> ***Hyperthermia* (or overheating of the body) deaths of children aren't confined to summer months.** They also happen during the spring and fall. Below are some examples.

- **Honolulu, Hawaii, March 7, 2007:** A three-year-old girl died when the father left her in a child seat for an hour and a half while he visited friends in a Makiki apartment building. The outside temperature was only 81°F.
- **North Augusta, South Carolina, April 2006:** A mother left her 15-month-old son in a car. He was there for nine hours while his mom went to work.

—Source: NOAA

THE SUN

Sunburn, with its ultraviolet radiation **burns, can significantly retard the skin's** ability to shed excess heat. In order to avoid sunburn, you need to be mindful of the sun's strength and intensity where you are. That depends on the weather. You may hear the term "UV Index" mentioned in your local weather report.

WHAT IS THE UV INDEX?
(Environmental Protection Agency)

Some exposure to sunlight can be enjoyable; however, too much could be dangerous. Overexposure to the sun's ultraviolet (UV) radiation can cause immediate effects, such as sunburn, and long-term problems, such as skin cancer and cataracts. **The UV Index, which was developed by the National Weather Service and the U.S. Environmental Protection Agency (EPA), provides important information to help you plan your outdoor activities to prevent overexposure to the sun's rays.**

The UV Index provides a daily forecast of the expected risk of overexposure to the sun. The Index predicts UV intensity levels on a scale of 1 to 11+, where 1 indicates a low risk of overexposure and 11+ signifies an extreme risk. Calculated on a next-day basis for every zip code across the United States, the UV Index takes into account clouds and other local conditions that affect the amount of UV radiation reaching the ground in different parts of the country.

UV Index number	Exposure level
2 or less	Low
3 to 5	Moderate
6 to 8	High
8 to 10	Very high
11 –	Extreme

The EPA's "SunWise" program recommends the following steps:

By taking a few simple precautions daily, you can greatly reduce your risk of sun-related illnesses. To be SunWise, consider taking the following action steps daily:

- Do not burn.

- Avoid sun tanning and tanning beds.
- Generously apply sunscreen with an SPF of at least 15.
- Wear protective clothing, including a hat, sunglasses, and full-length clothing.
- Seek shade.
- Use extra caution near water, snow, and sand.
- Watch for the UV Index.
- Get vitamin D safely.

Because children typically spend more time outdoors than adults, it is especially important that you make sure your children take these steps. Apply sunscreen to them (and teach them how to apply it to themselves when they are old enough). Also consider dressing them in protective wear, like hats and sunglasses when appropriate.

HOT WEATHER HEALTH EMERGENCIES
(CDC)

Even short periods of high temperatures can cause serious health problems. During hot weather health emergencies, keep informed by listening to local weather and news channels or contact local health departments for health and safety updates. Doing too much on a hot day, spending too much time in the sun, or staying too long in an overheated place can cause heat-related illnesses. Know the symptoms of heat disorders and overexposure to the sun, and be ready to give first-aid treatment.

HEAT STROKE
(CDC)

Heat stroke occurs when the body is unable to regulate its temperature. The body's temperature rises rapidly, the sweating mechanism fails, and the body is unable to cool down. Body temperature may rise to 106°F or higher within 10 to 15 minutes. Heat stroke can cause death or permanent disability if emergency treatment is not provided.

RECOGNIZING HEAT STROKE

Warning signs of heat stroke vary but may include the following:

- an extremely high body temperature (above 103°f, orally)
- red, hot, and dry skin (no sweating)
- rapid, strong pulse
- throbbing headache
- dizziness
- nausea
- confusion
- unconsciousness

WHAT TO DO

If you see any of these signs, you may be dealing with a life-threatening emergency. Have someone call for immediate medical assistance while you begin cooling the victim:

- Get the victim to a shady area.
- Cool the victim rapidly using whatever methods you can. Immerse the victim in a tub of cool water; place the person in a cool shower; spray the victim with cool water from a garden hose; sponge the person down; or wrap the victim in a cool, wet sheet and fan him or her vigorously.
- Monitor body temperature, and continue cooling efforts until the body temperature drops to 101–102°F.

- If the medical personnel are delayed, call the hospital emergency room for further instructions.

HEAT EXHAUSTION
(CDC)

Heat exhaustion is a milder form of heat-related illness that can develop after several days of exposure to high temperatures and inadequate or unbalanced replacement of fluids. It is the body's response to an excessive loss of water and electrolytes, the salt contained in sweat. Those most prone to heat exhaustion are elderly people, people with high blood pressure, and people working or exercising in a hot environment.

RECOGNIZING HEAT EXHAUSTION

Warning signs of heat exhaustion include the following:

- heavy sweating
- paleness
- muscle cramps
- tiredness
- weakness
- dizziness
- headache
- nausea or vomiting
- fainting

WHAT TO DO

Cooling measures that may be effective include the following:

- cool, nonalcoholic beverages
- rest
- cool shower, bath, or sponge bath
- an air-conditioned environment
- lightweight clothing

HEAT CRAMPS
(CDC)

Heat cramps usually affect people who sweat a lot during strenuous activity. This sweating depletes the body's salt and moisture. The low salt level in the muscles may be the cause of heat cramps. Heat cramps may also be a symptom of heat exhaustion.

RECOGNIZING HEAT CRAMPS

Heat cramps are muscle pains or spasms—usually in the abdomen, arms, or legs—that may occur in association with strenuous activity. If you have heart problems or are on a low-sodium diet, get medical attention for heat cramps.

WHAT TO DO

If medical attention is not necessary, take these steps:

- Stop all activity, and sit quietly in a cool place.
- Drink clear juice or a sports beverage.
- Do not return to strenuous activity for a few hours after the cramps subside, because further exertion may lead to heat exhaustion or heat stroke.
- Seek medical attention for heat cramps if they do not subside in one hour.

SUNBURN
(CDC)

Sunburn should be avoided because it damages the skin. Although the discomfort is usually minor and healing often occurs in about a week, a more severe sunburn may require medical attention.

RECOGNIZING SUNBURN

Symptoms of sunburn are well known: the skin becomes red, painful, and abnormally warm after sun exposure.

WHAT TO DO

Consult a doctor if the sunburn affects an infant younger than one year of age or if these symptoms are present:

- fever
- fluid-filled blisters
- severe pain

Also, remember these tips when treating sunburn:

- Avoid repeated sun exposure.
- Apply cold compresses or immerse the sunburned area in cool water.
- Apply moisturizing lotion to affected areas. Do not use salve, butter, or ointment.
- Do not break blisters.

(For more heat health information, go to www.bt.cdc.gov /disasters/extreme heat.)

PROTECT PETS IN HEAT

Dr. Charlotte Krugler and Dr. Adam Eichelberger, veterinarians from Clemson University in South Carolina, offer tips and answer questions on **how to protect your pets from extreme heat.**[3]

1. How do pets react in hot and humid conditions?
Dogs and cats often limit their activity on hot and humid days, which helps them stay cool. They can cool themselves by panting and also by "sweating" from their foot pads.

2. How do you know if it is "too hot" for your pet?
A pet owner should note how their pet normally breathes when lying quietly compared to light activity and vigorous exercise. Then they'll be more likely to notice abnormal clinical signs when a pet is too hot—such as heavy panting when not exercising, open-mouth breathing, inactivity, lack of appetite, mental dullness, and diarrhea.

When temperatures get hot, pets should not be encouraged to undergo vigorous activity for prolonged periods. Vigorous activity on a hot day increases the chances of a pet overheating, so owners should be especially vigilant during these times.

3. What precautions can pet owners take in the summer months?
- Never leave a pet unattended in a car on a warm or hot day.
- Keep fresh water available—cool when possible.
- Provide protection from direct sunlight; allow access to shade and shelter.

4. How can you take care of your pet in the heat?
NOTE: All animal owners are expected to provide adequate shelter and water to their animals. Pet owners should ensure that pets:

- are not confined in areas without air flow on hot and humid days
- are given a chance to acclimate to warmer temperatures. Examples:
  - If the owner and dog have been jogging together during the cooler months, the dog should go only a short distance on the first hot day, then gradually increase distance.
  - A dog that has been inside all winter and spring should not suddenly be put out in a hot yard for a ten-hour day. Rather, start with an hour at a time outside until the dog is used to the warmer temperatures.

WHAT TO DO IF YOU SUSPECT YOUR PET IS OVERHEATED

If pet overheating is suspected, the pet should be taken to a veterinarian immediately. In the interim, cool water (not warm, not cold) can be applied directly, or via dampened towels, to the pet's foot pads, to areas under armpits, inside back legs, and underside of the belly to help lower the pet's body temperature.

1. Can animals get sunburn? Should you observe the UV index?
Yes, animals can get sunburned (including dogs, cats, horses, pigs), so any bright day with direct exposure can result in a sunburn. Some animals seem to be more susceptible. Sunburned areas on white cats (usually ear tips) and pink-skinned areas on horses can progress to skin cancers.

Certain diets can induce photosensitivity, which will lead to sunburned skin in some animals.

2. Is heat tougher on certain breeds than others?
It seems that dog breeds with very thick, dense coats can have a tougher time in the heat. Some "arctic" breeds may be more suited to cooler climates. Also, black dogs are believed to absorb more heat into their bodies and seem to have problems on hot days, especially when exposed to direct sunlight.

3. What are common mistakes pet owners make in the warm-weather months?
The most serious mistake is to leave a pet in a closed automobile on a hot day. **Leaving the windows open will not prevent the temperature of the car, and then of the pet, from rising.** Many of these pets, even with rigorous treatment, do not survive, which is very sad since it is so preventable.

Another common mistake is over-exercising a dog on a hot day and not allowing the pet to cool down properly after a period of exercise. When dogs accompany owners on activities such as jogging, hiking, beach trips, etc., water should be provided at regular intervals and rest periods provided.

4. Can more frequent baths keep pets cooler?
Appropriate grooming and hygiene are recommended all year long, and will help the pet maintain a working cooling system. Here are more tips on bathing and grooming:

- If there is dirt buildup within the undercoat and hair, occasional bathing will help.

- Regularly combing out a dense, thick undercoat and removing mats will allow the skin to have access to air flow.
- Very thick-coated dog breeds may benefit from being shaved down by an experienced pet groomer once or twice during the hot season.

CONCLUSION

Spending too much time outside in extreme heat can be deadly. On average, each year heat kills more people in the US than hurricanes or tornadoes. Certain people are more vulnerable to extreme heat than others—like young children and the elderly. However, excessively hot days can be dangerous for everyone. It's advised to drink lots of cool water and take frequent breaks if you must spend time outside in extreme heat. Animals are also susceptible to heat-related illnesses. Pet owners should make sure that their animals have adequate shelter from the heat as well as fresh water accessible.

5
WILDFIRES

For legendary country singer Tanya Tucker, the move to Malibu, California, from Nashville, Tennessee, represented a fresh start.

"We wanted to be in Malibu because we love the ocean, and I thought my kids would love it out there, and we do," Tucker said.[1]

In summer of 2007, Tucker, along with her daughters, Presley and Layla, packed up their belongings and moved to the West Coast. They had their journey and transition to life in Malibu filmed for a new reality TV show they were developing.[2]

However, just a few short months later, in October 2007, Tanya and her family were faced with a terrifying new drama. This one wasn't for TV; it was happening in real life.

"Sunday morning I woke up. I got here to Las Vegas from Nashville on my way to L.A. and saw the news about the fires in Malibu. And right away I was worried about my eighteen-year-old daughter, Presley, who was in Malibu with our four dogs and all our animals. And that was my first concern—to get her out of the house." Tucker explained this by phone to HLN's Showbiz Tonight reporter Brooke Anderson on October 23, 2007.

Soon after Tucker's daughter Presley was evacuated, fire helicopters began dowsing the residence with water to keep the fire from spreading. "All my stage clothing, boots, belts, and wardrobe are in that house," Tucker says. "I have so much memorabilia since I just moved from Nashville to Malibu."[3]

Presley and the family's pets were evacuated from their home because of the nearby flames that were not yet contained. They were not the only ones. According to various news reports, over 1,500 people were forced to leave their homes due to the "Canyon Fire," as it was referred to by fire officials.

While many homes were spared or damaged in this fire, including Tucker's home (which various reports indicate was badly damaged by water and fire), other landmark properties were not. The fire destroyed *two famous historical structures in Malibu: the Castle Kashan and the Malibu Presbyterian Church.*

THE CANYON FIRE SWEPT THROUGH SOUTHERN CALIFORNIA

for six days. The cause of the fire was a mixture of several factors[4]:

- At the time the Canyon Fire developed, all of Southern California was experiencing an extreme drought. That year Southern California experienced one of the driest seasons on record.
- Relative humidity in the area dropped into the single digits, which is very low.
- The vegetation in the area, called chaparral, is extremely fire-prone. Because of the drought, the plants were brittle and dry, which made them more inclined to burn.
- Finally, the strongest factor that caused the Canyon Fire and others in the region like it to grow was the presence of the *Santa Ana winds*.

Santa Ana Winds are hot, dry winds that blow down from the mountains in Southern California. As the descending air heats and compresses, it can act as a giant hair dryer, heating tinder in advance of the fire and causing sparks and embers to jump forward. The particles can then ignite other trees and grass ahead of the main fire. While these winds can occur at any time of year, they tend to be more active in the fall and winter. During this time, high pressure over the high desert of the Great Basin region causes winds on the southern side of the high to blow from the east, toward the Pacific Ocean and lower air pressure offshore. The winds push dry air from the inland deserts of California and the Southwest over the mountains between coastal California and the deserts.

As the air descends from mountains, it is compressed and the temperatures increase. These hot, very dry winds (a relative humidity of 10–20 percent or lower is common) dry out vegetation, increasing the fuel available to feed fires. The gusty winds and eddies (or whirlpools of swirling winds) through canyons and valleys also fan flames and spread tinders.

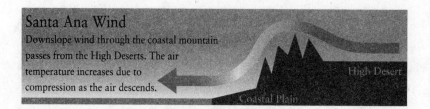

Santa Ana Wind

Downslope wind through the coastal mountain passes from the High Deserts. The air temperature increases due to compression as the air descends.

High Desert

Coastal Plain

WEATHER AND WILDFIRES

According to the National Interagency Fire Center, the last decade has seen an average annual wildfire-burned area of more than 6 million acres.[5] Michael Steinberg, Firewise Communities Program Manager of the National Fire Protection Association (NFPA), says, "Also, the trend of larger and more damaging fires is increasing, even though over the long run the number of fires every year has not varied that much. It's important not to understate the magnitude of this hazard." Wildfires can start out as small blazes and grow to enormous proportions. Carelessness is the chief cause of most wildfires. A stray cigarette or a campfire that was left not fully extinguished can have a devastating impact.

Wildfires can also be ignited by weather conditions. Lightning strikes are a common source of wildfires in the West.

> Lightning is the ignition for most wildfires in remote areas of the western United States. Dry thunderstorms with lightning and winds are major culprits in starting and spreading wildfires.
>
> Besides the weather brought in by regional cold fronts or local thunderstorms, wildfires can create their own weather. The heated air in a raging wildfire rises, sending water vapor released during combustion into the atmosphere. This buoyancy also produces intense updrafts and horizontal winds that shape and drive the fire line itself, possibly triggering sudden changes in direction and intensity that can threaten firefighters' lives.
>
> —The National Center of Atmospheric Research (NCAR) in Boulder, Colorado

"When it comes to fighting fires," says Don Whittemore, assistant chief of Rocky Mountain Fire in Boulder, Colorado, **"weather is either your best friend or your worst enemy.** It is the single greatest influence on fire behavior and, as such, directly determines one's strategies and tactics and ultimately the success."

" One of the most infamous fires of all time, the Yellowstone Fires of 1988, was completely dominated by weather patterns. The spring of 1988 began as a normal to above-normal moisture year. . . . By summer, however, the normal seasonal rain patterns dried up, average temperatures and seasonal winds increased, and the fires exploded. What fire behavior analysts experienced was unprecedented and contributed to a new level of understanding and appreciation of extreme fire behavior. **At the end of the season, it wasn't the thousands of firefighters or the millions of dollars who put the fires out; it was a quarter-inch of snowfall.**

— Don Whittemore, assistant chief of Rocky Mountain Fire

It might seem that the passage of a cold front is always beneficial for fire suppression. However, that's not always the case. While the front may bring a dousing of needed rain, it also can kick up the winds, spreading the fire further. Topography also comes into play when it comes to rapid fire growth.

" Flames at the base of a slope heat the vegetation higher up, releasing combustible gases that burst into new flames, and spread the fire uphill. The blaze moves up a hillside at speeds up to 15 miles per hour, but observers have seen acceleration as fast as 100 mph."

—National Center for Atmospheric Research

"FIRE WEATHER" FORECASTS
(NOAA)

"Computer models can forecast weather parameters that may lead to the possibility of a wildfire ten days out or longer. During fire season, every local National Wildfire Service office and fire agencies are in regular communication about the expected weather conditions for the next five to seven days. If there is any need for Fire Weather Watches or Red Flag Warnings, these issues are also discussed among the various agencies."[6]

You may hear advisories regarding fire weather issued on your local news or on your NOAA weather radio or local news broadcast. Here is what they mean:

Red Flag Warning: A term used by fire weather forecasters to call attention to limited weather conditions of particular importance that may result in extreme burning conditions. Its issuance denotes a high degree of confidence that weather and fuel conditions consistent with local Red Flag Event criteria will occur in 48 hours or less. Longer lead times are encouraged when confidence is very high or the fire danger situation is critical.

There is a rating system issued by the National Wildfire Service. On your local news or NOAA weather radio station you may hear that the danger of wildfire is *high* to *extreme* when the following forecast weather parameters are forecasted to be met:

- The sustained wind average is 15 mph or greater.
- Relative humidity is less than or equal to 25 percent and the temperature is greater than 75°F.
- In some states, dry lightning and unstable air are criteria.

Fire weather watch: Forecasters should issue a Fire Weather Watch when there is a high potential for the development of a Red Flag event. The watch will be issued 18 to 96 hours in advance of the expected onset of criteria. The watch may be issued for all, or selected portions within a fire weather zone or region. The overall intent of a Fire Weather Watch is to alert users at least a day in advance for purposes of resource allocation and firefighter safety.

TOP TEN STATES FOR WILDFIRES IN 2010

1. Texas
2. California
3. Georgia
4. North Carolina
5. New Jersey
6. Mississippi
7. Florida
8. Alabama
9. Louisiana
10. Arkansas

—Source: The National Interagency Coordination Center

If you are living in an area that's prone to wildfire, here are tips to protect your home: It's important to zone in on your home's "weakest links," according to the Los Angeles County Fire Department. They warn that **embers will find their way into your home through many hidden entry points.** These include:

Roofs: A roof is the most vulnerable surface for embers to land, lodge, and start a fire; this includes roof valleys, open ends of barrel tiles, and rain gutters.

Eaves: Embers gather under open eaves and ignite exposed wood or other combustible material.

Vents: Embers enter the attic or other concealed spaces and ignite combustible materials. Vents in eaves and cornices are particularly vulnerable, as are any unscreened vents.

Walls: Combustible siding or other combustible or overlapping materials provide a surface and crevice for embers to nestle and ignite.

Windows And Doors: Embers can enter gaps in doors, including garage doors. Plants or combustible storage near windows can be ignited from embers and generate heat that can break windows and/or melt combustible frames.

Balconies and Decks: Embers collect in or on combustible surfaces or undersides of decks and balconies, ignite the material, and enter the home through walls or windows.

" Perhaps the most fatal structural element is a wood roof. These are extremely flammable and their orientation and structure make them ideal recipients for embers. It's important to keep your gutters, vents, and other openings covered with fine mesh screening. I have seen leaves and conifer needles in these areas lead to fires. **During extreme fire events, I have also witnessed single-pane windows fail, resulting in the house eventually burning from the inside out.** Although not a weak point in the structure, many homeowners store woodpiles and other flammable objects such as lawn furniture adjacent to a house or underneath a deck or awning. These become good catch points for embers, and once they are on fire they start the structure on fire.

—Don Whittemore, assistant chief of Rocky Mountain Fire

Fire safety experts advise that you create a "defensible space" around your home. This means removing trees, shrubs, and grasses from near the house. This will make it harder for fires to start on your property.

There are several advantages of creating defensible space, according to Whittemore:

- By removing or reducing fuels around a structure, you are impeding the fire's ability to make contact with the building.

- You are creating a workable area where firefighters can employ tactics that have a chance of successfully defending the structure during a fire.

- **You increase your chances of salvaging property. When firefighters arrive on the scene they determine if it's "defendable"; a site that has an adequate defensible space will be more likely to receive additional efforts by firefighters than one that doesn't.**

The Colorado State Forestry Service has a checklist to go over to make sure your defensible space is at its best and most protective. As you survey your property, make sure:

- Trees and shrubs are properly thinned and pruned within the defensible space.

- Roof and gutters are clear of debris and that attic, roof, eaves, and foundation vents are screened and in good condition.

- Keep lawn mowers and agricultural equipment in proper working condition, and avoid rocks and other materials that might cause a spark.

- Branches overhanging the roof and chimney are removed and that chimney screens are in place and in good condition.

- An outdoor water supply is available, complete with a hose and nozzle that can reach all parts of the house.

- Fire extinguishers are checked and in working condition.

- The driveway is wide enough, and clear from excess tree branches, for fire and emergency equipment to get up to your house.

- There is an easily accessible tool storage area with rakes, hoes, axes, and shovels for use in case of fire.

- You have practiced family fire drills and your fire evacuation plan.

- The size of your defensible space depends on your property. The steepness of the slope that your home sits on is a factor to consider. Another is the trees, shrubs, and grass that grow around your home. If you are not sure how to build and clear this space around your home, your local fire department can help.

"People may neglect to maintain their defensible space from one year to the next. If keeping grass around a structure watered and cut is a component in a plan, then it must be maintained at a level consistent with the plan throughout the fire season. People 'forget' to actually do the necessary work. I know many fire-savvy people, including firefighters, who have lost their homes to a wild-fire because they 'forgot' to take care of the necessary business."[7]

FEMA offers more tips on wildfire preparations and precautions:

If you see a wildfire, call 911. Don't assume that someone else has already called. Describe the location of the fire; speak slowly and clearly; and answer any questions asked by the dispatcher.

- **First evacuate your pets and all family members who are not essential to preparing the home:** those with medical or physical limitations and the young and the elderly.
- **Wear protective clothing:** garments with long sleeves and pants that cover your skin.
- **Close and protect openings:** Close outside attic, eaves, and basement vents, windows, doors, pet doors, etc. Remove flammable drapes and curtains. Close all shutters, blinds, or heavy, noncombustible window coverings to reduce radiant heat.
- **Close inside doors and open damper:** Close all doors inside the house to prevent draft. Open the damper on your fireplace, but close the fireplace screen.

- **Shut off gas:** Shut off any natural gas, propane, or fuel oil supplies at the source.
- **Water:** Connect garden hoses. Fill any pools, hot tubs, garbage cans, tubs, or other large containers with water.
- **Pumps:** If you have gas-powered pumps for water, make sure they are fueled and ready.
- **Ladder:** Place a ladder against the house in clear view.
- **Car:** Back your car into the driveway and roll up the windows.
- **Garage doors:** Disconnect any automatic garage door openers so that doors can still be opened by hand if the power goes out. Close all garage doors.
- **Valuables:** Place valuable papers, mementos, and anything you can't live without inside the car in the garage, ready for quick departure.
- **Leave a note for firefighters about any special hazards on the property.**
- **Cut or weed whack any grasses surrounding your home.**
- **Run any sprinklers to wet down the vegetation.** (According to Whittemore, "Watering or using your sprinklers on the roof is generally a waste of time.")

PREPARING TO LEAVE

The Los Angeles County Fire Department advises that it's best to leave "early enough to avoid being caught in fire, smoke, or road congestion. Don't wait to be told by authorities to leave. In an intense wildfire, they may not have time to knock on every door. If you are advised to leave, don't hesitate!"

If you have to evacuate, make sure you have your Family Disaster Plan in place and ready to execute. Details on how to create this plan can be found in Chapter 13.

Don Whittemore of the Rocky Mountain Fire Department in Boulder, Colorado, has more guidance on evacuation in case of wildfire. He advises to **write out your evacuation plan and put items in priority order.** "No matter what one's list contains, the key is that it is prioritized based on an individual's values and made with the understanding that there may be instances when there is only enough time to do some of the tasks on the list and not all."

- Practice the plan ahead of time.
- Update it annually, so everyone in the household is familiar with what to do.
- If any family members have special needs, that should be taken into consideration.
- Time it out. It's also crucial to rehearse how you would react in this situation and to time how long each action takes.

After fighting fires for over 30 years and knocking on hundreds of people's doors to enforce evacuations, Whittemore says there is no way to know how much advance notice you will get from authorities.

I have heard tales from several people, when under the stress of having only fifteen or so minutes to evacuate, they end up taking or doing nothing because they wasted those precious minutes trying to figure out what was most important. . . . **By timing your plan you'll have better and more realistic expectations of what you can accomplish given various time intervals.**

In some cases I've given folks twenty-four hours' notice and in other instances it's been a "right now" situation.

FEMA adds that if you have to evacuate due to the threat of wildfire, other precautions to take include:

- **Lights:** Turn on outside lights and leave a light on in every room to make the house more visible in heavy smoke.

- **Don't lock up:** Leave doors and windows closed but unlocked. It may be necessary for firefighters to gain quick entry into your home to fight fire. The entire area will be isolated and patrolled by sheriff's deputies or police.

WHAT TO DO DURING A WILDFIRE IF YOU ARE AT HOME

- In the worst-case scenario, where you are trapped at home and have already called 911, FEMA advises that you "stay inside and away from outside walls. Close doors, but leave them unlocked. Keep your entire family together and remain calm."

The Los Angeles County Fire Department offers these additional survival tips:

- Patrol inside your home for spot fires and extinguish them.
- Wear long sleeves and long pants made of natural fibers, such as cotton.
- Stay hydrated.
- Ensure that you can exit the home if it catches fire (remember, if it's hot inside the house it is four to five times hotter outside).
- After the fire has passed, check your roof and extinguish any fires, sparks, or embers.
- Check inside the attic for hidden embers.
- Patrol your property and extinguish small fires.

WHAT TO DO DURING A WILDFIRE IF YOU ARE TRAPPED IN YOUR CAR

- This is dangerous and should be done only in an emergency, but you can survive the firestorm if you stay in your car, according to FEMA. Most people don't know that metal gas tanks and containers rarely explode. It is much less dangerous than trying to run from a fire on foot.

- Roll up windows and close air vents. Drive slowly with headlights on. Watch for other vehicles and pedestrians. Do not drive through heavy smoke.
- If you have to stop, park away from the heaviest trees and brush. Turn headlights on and ignition off. Roll up windows and close air vents.
- Get on the floor and cover up with a blanket or coat.
- Stay in the vehicle until the main fire passes.
- Stay in the car. Do not run! Engine may stall and not restart. Air currents may rock the car. Some smoke and sparks may enter the vehicle. Temperature inside will increase.[8]

WHAT TO DO DURING A WILDFIRE IF YOU ARE CAUGHT ON FOOT

- The best temporary shelter is away from any vegetation. On a steep mountainside, avoid canyons.
- If a road is nearby, lie facedown along the road in the ditch. Cover yourself with anything that will shield you from the fire's heat.
- If hiking in the back country, seek a depression with sparse fuel. Clear fuel away from the area while the fire is approaching and then lie facedown in the depression and cover yourself. Stay down until after the fire passes.

WHAT TO DO AFTER A WILDFIRE

- Check the roof immediately. Put out any roof fires, sparks, or embers. Check the attic for hidden burning sparks.
- The water you put into your pool or hot tub and other containers will come in handy now. If the power is out, try connecting a hose to the outlet on your water heater.
- For several hours after the fire, maintain a "fire watch." Recheck for smoke and sparks throughout the house.[9]

HOW TO PROTECT YOUR PETS IN A WILDFIRE

Colleen Paige is a family and pet lifestyle expert, an animal behavior specialist, and a former paramedic. According to her website, she has had the personal experience of evacuating families during wildfires. "Many evacuations I coordinated as a paramedic gave a fifteen-minute window only," she writes on her website, ColleenPaige.com. "Cat owners suffered the most, with outdoor cats they couldn't find. Many were forced to leave them behind and hope for the best."[10]

EVACUATING WITH YOUR ANIMALS

Paige says:

> If you must vacate with a dog or cat, just before fleeing from home, attempt to wet your pet down and cover him or her with a wet towel to prevent flying embers from burning their fur and paws on the way out of the house. Also, be sure to hose down the ground from house to car to further protect exposed paws.[11]
>
> **If you must evacuate and can't find your pet, leave a door open and call the pet's name once you get out. Dogs respond better to their owners' voices than cats. If you can't find them, make sure to leave an outside door open.**
>
> **If you must vacate with a horse or farm animal,** hose down the animal to prevent burns due to flying embers. Cover the animal's eyes with blinkers and lead into a trailer. Make sure to load in a secure area, keeping all corral gates and fences closed to prevent the animal from fear-bolting and running off.
>
> This will be a frightening time for both people and pets. The aftermath will be far-reaching, and it will take many months for people, pets, and wildlife to begin to recover from the physical and emotional

trauma. Preparation and planning can make the critical difference in protecting loved ones and saving lives.[12]

If you have pets you have to take them into consideration when planning for an emergency. Find out how to create a Family Disaster Plan for People and Pets in Chapter 13. Items needed for a Pet Emergency Supply Kit are detailed in Chapter 15.

Here are more guidelines, from petinsurance.com (Veterinary Pet Insurance—A Nationwide Company) and the Humane Society, on how to care for pets in the event of a wildfire:

Alert Firefighters to Indoor Pets

Place a Pet Rescue Fire Safety Sticker in Your Window: Fires often break out when owners are out and these stickers have saved many a pet's life. You can search online nonprofit humane organizations to request a free pet emergency sticker. These stickers are available from most pet stores and nonprofit humane organizations. "They let fire crews know that you have pets inside the house, how many, and what kind."[13]

Write down the number of pets inside your house and attach the sticker to a front window where it can be easily found by emergency responders. Include special instructions, like their favorite hiding places. This critical information saves rescuers time when locating your pets. You can obtain a free window cling by going to www. adt.com/pets or from the AKC (American Kennel Club), http://www.akc.org/ and the ASPCA, http://www.aspca .org/.

"Pets Left Alone Can't Escape a Burning Home: Consider using monitored smoke detectors that are connected to an emergency response center with people on hand who can call the fire department in your absence.

Keep Outdoor Pets Away from Danger: Keep pet houses or pens away from brushy areas. Fire departments will warn you to clear dry brush away from your home, but that also applies to your pets."[14]

"If you have a doghouse or a pen for a rabbit, potbellied pig, or other outdoor pet, make sure it's at least 20 feet away from any brush that could possibly become fuel in a fire. That way, you'll have time to go out and rescue your pet if such a fire does threaten your property."[15]

CONCLUSION

Wildfires are mainly caused by human negligence. However, weather events like lightning strikes can also ignite the flames. Strong winds, like those in California called Santa Anas, can cause rapid fire growth. Residing in a particular geographic area may put your home at greater risk. There are precautions you can take, such as clearing excess branches around your property, that can help protect your home when a wildfire threatens. When that happens, there is often little time to evacuate. This is why you should have your evacuation plan ready in advance and rehearse it. This is the best way to ensure that you get your family and your pets to safety.

6
RIP CURRENTS

"I would never have dreamed in a million years that you could be onshore . . . standing in waves and a current can knock you off your feet and take you out. Yet, that's what happened to Larry. . . . It should never have happened."[1]

Sandee LaMotte speaks in a soft, but steady voice. She tells the story of her husband's death slowly and purposefully. Sandee lost her husband, Larry LaMotte, on Sunday, June 8, 2003.

That weekend was to begin their long-anticipated summer vacation. Like many people who lived in Atlanta, they chose Florida's Panhandle beaches as a vacation spot.

This particular year, the vacation came at an exciting time for the LaMotte family. Larry, who had once been a bureau chief and correspondent for CNN, was embarking on a new career as a life coach. Sandee was doing well, too. She was on the road to advancement in her job at WebMD, a website based in Atlanta. Sandee and Larry's two children, then 9-year-old Krysta and 12-year-old Ryan, were in good spirits as they loved to go to Grayton Beach.

The family arrived in Florida on Sunday afternoon. On their walk down to the beach from the cottage they rented, Sandee saw a small sign with a red flag in the sand that indicated there could be strong currents in the water.

The kids, who'd been splashing in a nearby lake earlier, were looking forward to jumping in the ocean with their boogie boards. There was no lifeguard on duty. Since there were so many others in the water, Sandee allowed the children to go in the ocean, as long as they stayed close to the shore. Her understanding of rip currents at the time was that they were a danger to people who swam well away from the shoreline, not just a few feet from the sand.

The kids were having fun in the waves. It was almost 6 p.m. now, and Sandee told Larry she'd go back to the house to get dinner started. Shortly after Sandee left, Ryan called out to Larry, who was *standing on the beach, that he couldn't swim back to shore. Ryan was having trouble maneuvering his boogie board back in the rough waves. The boy was struggling, so Larry went into the water to help him. Witnesses say Larry disappeared right after entering the water. They describe seeing him one instant and gone the next. They estimate he was taken out 100 yards by the rip current. A man saw what was happening, and he yelled to Ryan to paddle sideways. Ryan followed the man's instructions, and soon the boy was back on the beach. But where was Larry?*

The brother-in-law of the man who coached Ryan sprang into action. He ran into the water and swam out to save Larry. His name was Ken Brindley, and, like Larry, he was also the father of two young children.

Ryan and Krysta ran back into the house to tell their mom what was happening. "I left first, to get a head start on dinner. Ten minutes later Ryan and Krysta came bursting through the door. 'Mom, come quick. Ryan was stuck in the water and Daddy went in after him. Now Daddy's gone!'" Sandee recalls.

Sandee's voice gets softer, and she slows down as she describes what happened next.

"Well, when I was running toward the beach I could see helicopters flying in and I could see all these people and I just ran straight into the water and stood there and I think someone thought I was going to swim out there . . . so they grabbed me and held me back. I remember looking down and seeing Larry bob in the water and there was no rope that came down from the helicopter, no inner tube to float him in. I just screamed. I just screamed. . . . All I could think was: Oh my God, my son Ryan is going to blame himself, my son is going to blame himself. . . ."

Sandee's eyes close, and her voice trails off. Larry's body, covered in sand, was pulled out of the water. Sandee knew he was dead when she saw him. He was taken to the hospital by ambulance and pronounced dead there. Larry was 60 years old.

Ken Brindley, 36, of Conway, Arkansas, died in the hospital two days later. Ken left behind his wife, Melanie, and two children: three-year-old Blake and six-year-old Madeline. Ken didn't know Larry. Ken was simply a Good Samaritan who made the choice in an instant to attempt to save the life of a man he never met before.[2]

A total of nine people drowned that weekend on the waters along the Florida Panhandle.[3]

EPILOGUE

Sandee LaMotte is working toward using the tragedy of Larry's and Ken Brindley's death to educate others on beach safety. She is working with dedicated individuals to educate elementary school children on water safety. She is also working on a bill in Congress to establish "Larry's Law," which would require lifeguards on all public beaches.

WHAT IS A "RIP CURRENT"?

Rip currents are powerful, channeled currents of water flowing away from shore. Rip currents can be as narrow as 10 or 20 feet in width, though some may be up to ten times wider. They can extend for hundreds of feet beyond the surf zone, the area between the high-tide level on the beach to the seaward side of the breaking waves. They typically extend from the shoreline, through the surf zone, and past the line of breaking waves. Rip currents can occur at any beach with breaking waves, including the Great Lakes. Rip currents most typically form at low spots or breaks in sandbars, and also near structures such as groins, jetties, and piers.[4] *Rip currents* are not *rip tides*. Even though the terms are mistakenly used interchangeably, they are caused by different phenomena. The correct term to use is rip currents, according to the NOAA and the USLA United States Lifesaving Association.

HOW DOES THE FLOW OF A RIP CURRENT MOVE?

When you compare normal ocean waves to those with rip currents, you can see the rip current pulls away from the shore, back out to sea.

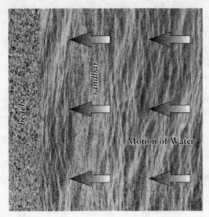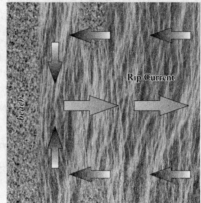

Notice that the flow is horizontal, not vertical. Rip currents don't pull you under. They pull you out to sea.

Rip current speeds can vary. Sometimes they are too slow to be considered dangerous. However, under certain wave, tide, and beach conditions, the speeds can quickly become dangerous. Rip currents have been measured to exceed 5 mph, slower than you can run but **faster than you or even an Olympic swimmer can swim.** In some cases they have been measured as fast as 8 feet per second. This is faster than the speed at which an Olympic swimmer can swim a 50-meter sprint.[5]

Rip currents can be killers. **The United States Lifesaving Association estimates that the annual number of deaths due to rip currents on our nation's beaches exceeds 100.** Rip currents account for over 80 percent of rescues performed by surf beach lifeguards, according to the United States Lifesaving Association.

The greatest safety precaution that can be taken is to recognize the danger of rip currents and always remember to swim at beaches with lifeguards. The United States Lifesaving Association has calculated the chance that a person will drown while attending a beach protected by USLA-affiliated lifeguards at 1 in 18 million.

"**Stay out of the water if you are at all uncertain. Never turn your back on the sea, since waves come in varying sizes and you can easily be knocked over and dragged out. Have your kids wear Coast Guard–approved personal floatation devices,**" says B. Chris Brewster, president of the United States Lifesaving Association.[6]

WHAT IS UNDERTOW?

Undertow, an often misunderstood term, refers to the backwash of a wave along the sandy bottom. After a wave breaks and runs up the beach face, some of the water percolates into the sand, but much of it flows back down the beach face, creating a thin layer of offshore-moving water with a relatively high velocity. This backwash can trip small children and carry them seaward. However, the next incoming wave causes higher landward velocities, pushing them back up on the beach. **Undertow is not a rip current. Undertow does not pull you underwater or out to sea.**[7]

HOW CAN YOU KNOW IF RIP CURRENTS ARE IN THE OCEAN?
(NOAA)

Signs that a rip current is present are very subtle and difficult for the average beachgoer to identify. Look for differences in the water color, water motion, incoming wave shape, or breaking point. Look for any of these clues:

- channel of churning, choppy water

- area having a notable difference in water color
- line of foam, seaweed, or debris moving steadily seaward
- break in the incoming wave pattern
- One, all, or none of the clues may be visible.

Note that all these factors don't have to be occurring for rip currents to be present. You may notice just one or two of them. The USLA advises that you use polarized sunglasses, which can help reduce glare from sunlight reflected on the water, at the beach to look for these characteristics more clearly.

FORECASTS AND ADVISORIES

Issuing rip current advisories or warnings is within the given jurisdiction of the local beach patrol, local lifeguards, or local law enforcement officials.

NOAA does mention risk levels of rip currents in its Hazardous Weather Outlook (HWO). That information can also be found in NOAA's forecasts for the "surf zone" of many waterfront communities. Many NWS forecast offices issue a surf zone forecast during the traditional summer season from Memorial Day through Labor Day. The surf zone forecast typically describes the following parameters and hazards: sky condition, precipitation, visibility, air temperature, wind speed and direction, wave height, surf temperature, tide information, **rip currents**, lightning, severe thunderstorms, and the ultraviolet index. It provides valuable and lifesaving information on the hazards of the surf zone to the greater beachfront community, to include the general public and the providers of beachfront safety services, such as lifeguards. It's a good idea to look this up before heading to the beach.

NATIONAL WEATHER SERVICE OFFICES
ISSUING SURF ZONE FORECASTS

Atlantic	Gulf Coast
Boston, MA	Tallahassee, FL
Miami, FL	Tampa, FL
Melbourne, FL	Mobile, AL
Jacksonville, FL	Brownsville, TX
Charleston, SC	Corpus Christi, TX
Wilmington, NC	
Morehead City/Newport, NC	**Pacific Coast**
Upton, (NYC) NY	Los Angeles, CA
Wakefield (Norfolk), VA	San Diego, CA
Philadelphia/New Jersey	Honolulu, HI
	Guam
Great Lakes	
Duluth, MN	**Alaska**
Chicago, IL	Barrow
Northern Indiana, IN	
Gaylord, MI (starting July 1, 2010)	
Grand Rapids, MI	
Marquette, MI	

Not all coastal and Great Lakes NWS Forecast Offices provide a Surf Zone Forecast; however, **all NWS offices forecasting moderate or high risk of rip currents include this information in the Day 1 portion of their Hazardous Weather Outlook.**

Surf Zone Forecasts will contain Rip Current Outlooks. Rip Current Outlooks use the following, three-tiered set of qualifiers:

- **Low Risk** of rip currents. Wind and/or wave conditions are not expected to support the development of rip currents; however, rip currents can sometimes occur, especially in the vicinity of groins, jetties, and piers. Know how to swim and heed the advice of lifeguards.

- **Moderate Risk** of rip currents. Wind and/or wave conditions support stronger or more frequent rip currents. Only experienced surf swimmers should enter the water.
- **High Risk** of rip currents. Wind and/or wave conditions support dangerous rip currents. Rip currents are life-threatening to anyone entering the surf.

NOAA advises: "When you arrive at the beach, speak with on-duty lifeguards about rip currents and all other water conditions expected for the day."

> ## HURRICANES, TROPICAL STORMS, STRONG AREAS OF LOW PRESSURE CAN ALL CAUSE RIP CURRENTS TO FORM, BEFORE, DURING, AND AFTER THE STORM
>
> When a storm system produces strong winds over ocean waters, it usually causes the water to rise near center of the storm. As these waves move away from the storm, they can grow in height to larger swells. These swells push more water toward the coastline. All this extra water piles up and as it subsides seaward, the greater weight of water from the heavy waves that piled on shore can carve out trenches in sandbars within the surf zone. **As this large mass of water moves back out to sea in these newly formed channels or "rips," it creates a strong seaward retreating force or current of water. That is how the term "rip current" gets its name.**
>
> —Source: NWS, Jacksonville, FL

FLAGS AT THE BEACH

Keep in mind when you are at the beach to look for signs and flags for advisories. Sandee LaMotte recalls seeing a small sign with a red flag saying STRONG CURRENTS COULD BE IN THE WATER, and there was no lifeguard.

Today, Florida requires the signs to uniformly explain what each color means on all state beaches. "The Florida Legislature decided that a uniform flag system would provide the best measure of safety."[8]

According to the Florida Department of Environmental Protection, "The beach flags provide general warnings about overall surf conditions and do not specifically advise the public of the presence of rip currents."[9]

B. Chris Brewster, the president of the United States Lifesaving Association, worked with the state of Florida to develop the flag code sign. He recommends that:

- Flags should be used only on beaches with lifeguards present.
- **"If there are no lifeguards, we recommend not swimming in the ocean."[10]**

RIP CURRENT SAFETY

Additional Tips from NOAA:
#1 Learn how to swim!
When at the beach:

- Whenever possible, **swim at a lifeguard-protected beach.**
- Never swim alone.
- Learn how to swim in the surf. It's not the same as swimming in a pool or lake.
- Be cautious at all times, especially when swimming at unguarded beaches. If in doubt, don't go out.
- Obey all instructions and orders from lifeguards. Lifeguards are trained to identify potential hazards. Ask a lifeguard about the conditions before entering the water.

- Stay at least 100 feet away from piers and jetties. Permanent rip currents often exist alongside these structures.
- Consider wearing polarized sunglasses when at the beach. They will help you spot signatures of rip currents by cutting down glare and reflected sunlight off the ocean's surface.
- **Pay close attention to children and the elderly when at the beach.** Even in shallow water, wave action can cause loss of footing.

WHAT TO DO IF YOU ARE CAUGHT IN A RIP CURRENT

- Remain calm to conserve energy and think clearly.
- **Never fight against the current.**

SURVIVING A RIP CURRENT

" The mistake people make is to fight the current. Imagine trying to stop the power of an incoming wave. Can you do that? For the same reason, very few people can fight the outgoing power of a rip current. **The key to surviving in the surf zone is to respect the power of the ocean and to realize you can't win by fighting it directly.** You want to use the power to your advantage, like swimming out of the rip current and then using the power of the waves to help push you to shore.

—B. Chris Brewster, USLA

- Swim out of the current in a direction following (think parallel to) the shoreline. When out of the current, swim at an angle— away from the current—toward shore.

RIP CURRENTS
Break the Grip of the Rip!

ESCAPE   ESCAPE

ESCAPE   ESCAPE

CURRENT   **RIP CURRENT**   CURRENT

Rip currents are powerful currents of water moving away from shore.
They can sweep even the strongest swimmer out to sea.

Here's why you should swim parallel to the shoreline, according to the USLA:

- Most rip currents pull away from shore in a perpendicular or angled direction. They tend to narrow as they move away from shore, then to disperse at varying distances from shore.
- Swimming against the rip current will cause most people to quickly tire and panic.
- Swimming to the side, parallel to shore, often allows one to escape the current. If that doesn't work, one can float with the current and just tread water, until it disperses.
- Although people may find it scary to allow themselves to be carried away from shore, most people can tread water for a long time, but they can fight the rip for only a short time before exhausting themselves.
- If you are still unable to reach shore, draw attention to yourself by waving your arms and yelling for help.

In the tragic deaths of Larry LaMotte and Ken Brindley, both men drowned while trying to save someone else struggling in a rip current. This is not unusual, according to Dr. Peter Wernicki, medical advisor to the U.S. and World Lifesaving Associations: **"It's often not the first person caught in the rip current, but the second or third, who drowns. The child who might be a bit more buoyant comes out okay. But the dad who charges right in, he is the one who often succumbs.** I think essentially it's a matter of exhaustion. People fight and fight and fight and start swallowing water and just go below the waves."[11]

WHAT TO DO IF YOU SEE SOMEONE CAUGHT IN A RIP CURRENT

- Get help from a lifeguard.
- If a lifeguard is not available, have someone call 911.
- Throw the rip current victim something that floats—a lifejacket, a cooler, an inflatable ball.
- Yell loudly and clearly instructions on how to escape:
  1. Don't swim against the current.
  2. Swim parallel to shore in either direction.
  3. (If that doesn't work) relax and tread water.
- Remember, **many people drown while trying to save someone else from a rip current.**

> "See if you can throw them something that floats or put some floatation device in the rip," advises B. Chris Brewster, who is considered to be one of the most experienced and knowledgeable lifesaving water rescue experts in the world. "Don't try a rescue unless you are very experienced at water rescue, and always take some sort of floatation device."

CONCLUSION

Rip currents can occur on any beach, even one that seems safe and is filled with people. As you plan your beach vacation, educate yourself about the beach's history. Have there been fatalities due to rip currents? Learn the telltale signs and remember the basics of what to do if you are caught in a rip current: remain calm and swim parallel to the shore until you are out of the rip and can safely swim back to the shore. Learn the lifeguard schedule of the beaches you frequent, and do not swim if there are no lifeguards on duty.

7
LANDSLIDES AND MUDSLIDES

She's a grown woman today, but not too long ago Nicole James[1] was an 18-year-old girl who was nearly killed when her beach house became a deathtrap in a catastrophic landslide.

In the days preceding the natural disaster in La Conchita, California, the weather had been tumultuous. Rain had been pouring down in torrents in this a beachfront community midway between Ventura and Santa Barbara. On Sunday, January 9, 2005, there were strong thunderstorms off and on. That night a loud clap of thunder woke Nicole up from her sleep. A few hours later, it was early Monday morning and time to get up. Nicole dressed and headed off to school. It was the first day back from winter break at her college in Santa Barbara, and she didn't want to be late.[2]

She left the beach house she was renting from a family friend in La Conchita, and soon she was driving on Highway 101 headed north to Santa Barbara. Traffic was barely crawling due to the rain. Then it came to a complete halt. Nicole would soon learn that mudflow from two points, north and south of her, caused a backup of over 200 cars stuck in between. There was nowhere she could go.

Another stranded driver—headed in the opposite direction along the 101, however—was Nicole's new neighbor, Bill Harbison.[3] The two had never met before and didn't interact at that time, either. Neither of them knew then that their meeting would occur only hours later in the most dire circumstances.

Nicole was stuck in traffic for hours. Finally, officials tapped on Nicole's window and told her she'd have to abandon her car and leave a note with her name, number, and the keys inside.

She walked a mile back home from the drop-off point in La Conchita. It was late afternoon by now. Nicole entered the beach house and went upstairs to the living room. She sat on the sofa, waiting for a phone call from her boyfriend. Then something strange happened. Nicole remembers:

*The walls began to move. I thought it might be an earthquake
at the time. Except instead of swaying back and forth—like
they normally would in an earthquake—the walls just seemed
to be buckling in one direction. I grabbed my cell phone on
the coffee table and moved quickly toward the spiral staircase
to go downstairs. As soon as I got to the top of the stairs, I
felt dirt falling from the ceiling on my face. Some of it went
into my eye, and that was as far as I got.*

In the minutes that followed, the beach house collapsed on itself,
with Nicole trapped inside:

*Something inside me knew what was happening. I covered
my head with my arms. . . . At the top of the stairs there was a
huge cabinet. . . . It pushed me down, but it also protected me
in a small space. I heard things breaking and crashing. I tried
to move, but the only thing I could move was my left foot.
I knew I needed to scream or something so someone would
hear me and help me. So that's what I did.*

Earlier that morning, when mudslides at two points on Highway
101 trapped morning commuters, one of the people stuck was Nicole's
neighbor, 35-year-old Bill Harbison, a nurse who was driving home
to La Conchita after working a double shift at a hospital in Santa
Barbara. He had been looking forward to getting some rest in his new
"dream home" that he'd purchased only a month and a half before.

The 18-mile drive from the hospital had been tough. It was the
worst weather he'd encountered since moving there. "Thunder, lightning, nonstop heavy rain. It was like a hurricane," he remembers.

As Bill got closer to La Conchita, he could see the mudflow in
front of him that had stopped traffic. His windshield wipers swished
across his windshield at high speed, trying to clear the heavy rainfall.
Through the glass, he could make out the hillside in the distance. "I
could see the mud flowing down. Sometimes the mud would stop.
Then, every now and again, something would move on the hillside,
and more mud would rush down," he recalls.

Six hours later, when the weather let up a bit, Bill was able to get
his vehicle through the one lane that became open. Once home in La
Conchita, instead of sleeping, he wanted to see how his new neighborhood was faring in this weather.

Bill thought the best way to do this would be by bicycle. He opened the front door, and that's when he heard the sound:

> *It sounded like a pop, and then a low rumble. I looked to the right and the hillside was literally coming down, like a freight train, on top of the town. I could hear metal bending and branches breaking. Dust and debris were flying up and everywhere. Telephone poles were toppling over. Wires snapped. I watched as the mudslide stopped in front of me, just one block away from my house. . . . I hopped on my bike and started riding straight up to it. I got off and started calling out to people I thought may be trapped.*

Before Bill became a nurse, he had been an EMT, so it was his instinct to rush in and help. He hopped on his bicycle and pedaled quickly to the street that looked like it was hit the hardest. Once he got there, he jumped off his bike and went into search-and-rescue mode. "Hello!" he called out to piles of rubble that had been standing homes moments earlier. "Can anybody hear me?"

The responses came, but they were muffled. People were buried alive, and because many people called out to Bill at the same time, he had trouble pinpointing their location.

As he pulled apart debris, Bill climbed through piles of dirt, stones, and wood, scraping his legs and hands on items like mini-blinds and broken light fixtures.

The fast-moving mud that had flowed into the town had solidified. It not only knocked down homes, but it encased their remains with solid mounds of earth. It was hard enough to support Bill's weight; he was able to stand on top of it without sinking. Standing atop the collapsed homes now in ruins, he looked down at his surroundings to get a sense of where he might be in someone's house. All of the houses on this street had fallen down in heaps. Spaces that were once side yards that separated homes from one another were covered by mounds of solid mud and debris. It was not clear where one house's rubble ended and another's began. They seemed to blend into each other. Was he above a kitchen? A bedroom? It was difficult to tell as the homes now lay in dirt-filled ruins.

He called out again. Bill knew there were people trapped and he was determined to get them out. Through the muffled voices, the one Bill could hear crying out for help the clearest was Nicole's.

Following the sound of her voice, he dug through drywall, furniture—whatever he could to get to her. It worked. He managed to move enough debris off her to pinpoint where exactly she was. "I could see what looked like a teeny bit of her foot or leg sticking out," Bill remembers.

Nicole was pinned between a large bookcase and the stairs. Bill tried to keep her calm. "Hi, I'm Bill," he told her. "I'm going to get you out of here no matter what."

At that point a Ventura City police officer showed up. He called out to Bill: "What do you need? Do you need tools? I'll get them for you."

The cop left and returned shortly with a sturdy shovel that allowed Bill to dig deeper in order to get to Nicole. "She was in this cocoon between the furniture. She was so lucky because where she was allowed her not to be crushed. It held just enough space around her to keep her safe."

Bill's nursing skills came into play right away. "Are you having any trouble breathing?" he asked her.

"A little," she told him. "I think I cut my leg."

He looked down at the deep gash on her leg. "It's not too bad." He wanted to reassure her. "You're going to be fine."

Bill eventually carried Nicole to an ambulance, and she was on her way to a hospital. Her injuries were not life-threatening, and she recovered at her father's house in Santa Clarita.

In the weeks that followed, Nicole had the chance to thank Bill in person for saving her life. Her father, Luis, came with her to thank him as well.

Three years later, in 2008, Nicole got married, and she invited Bill. The wedding was a joyous occasion filled with music and dancing. Getting ready to leave the party, Bill tapped Nicole on the shoulder to say goodbye and congratulate her and her new husband. A huge smile spread over the young woman's face upon seeing her hero. She took Bill's hand and announced, "This is him! This is the man who saved my life!"

Family and friends gathered around to meet Bill and shake his hand. Nicole's grandmother took Bill's hands in hers and looked into his eyes. It's a moment Bill says he will never forget. "Thank you! Thank you for saving our Nicole!" she told him. "It is because of you she is here today!"

According to the Los Angeles Times, *Nicole "was one of 13 people rescued from the mud that killed 10 people . . . at least 400,000 tons of mud and debris collapsed from the 600 foot cliff that borders the town to the east.*"[4]

MOST OF THE HOMES IN LA CONCHITA SIT ON A NARROW
strip of land that's about 800 feet from the shoreline and a 600-foot-high bluff. The slope is steep, at around a 35-degree angle. Dr. Randall Jibson of the USGS (United States Geological Survey) Landslide Hazards Reduction Program says the landslide occurred because of the presence of a steep, high bluff composed of very weak sediment.[5] The quality of the rock and earth in the slope make landslides possible. The heavy rain that had been occurring in the weeks before made the situation more dangerous. The 2005 landslide happened at the end of a 15-day period that produced record and near-record amounts of rainfall in many areas of Southern California.

MUDSLIDES VS. LANDSLIDES

The terms "mudslides" and "landslides" are often used interchangeably to describe rapid, destructive debris flow. The difference is in the type and size of debris. However, there is some confusion on the definition of "landslides," "mudslides," and "debris flow." Dr. Randall Gibson of the United States Geological Survey explains:

> **Landslide** is a generic term for all downslope movement of earth materials. Thus, for example, rock falls and debris flows are types of landslides and could appropriately be called either by their technical names or simply referred to as landslides. **Debris flows** are landslides that consist of a mixture of soil, rock and water that flow in a semi-liquid state and can move large distances at high velocities and thus be very destructive. **Mudslide** is a popular term that has no technical meaning. In other words, mudslide is a term that should not be used because it has no technical definition. Also, mud doesn't slide, it flows, and so this term is misleading.

Dr. Jonathan Stock, also of the USGS, adds: "Mudslide is an ambiguous term for debris flows *used by the media,* often with the implication that the debris flow have detectable amounts of silt or mud. . . . Most of us no longer use it *except* for debris flows that are almost entirely silt and mud and lack boulders . . ."

LANDSLIDES CAN ALSO BE CAUSED BY VOLCANOES

Large landslides have swept down the slopes of Mount Rainier, Washington. The largest volcano landslide in historical time occurred at the start of the May 18, 1980, Mount St. Helens eruption.

—USGS

Landslides can occur at any time of year. They are more likely, though, in times of heavy rainfall. In some areas, that may be the spring or summer months. It's important to be aware of where you live in terms of proximity to a steep hillside and the times of year your region receives the most rain. Also, be aware of your region's susceptibility to earthquakes, since they can precipitate avalanches.

AREAS WHERE LANDSLIDES OCCUR IN THE UNITED STATES

Landslides occur in every state and U.S. territory. The Appalachian Mountains, the Rocky Mountains, and the Pacific Coastal Ranges and some parts of Alaska and Hawaii have severe landslide problems. Any area composed of very weak or fractured materials resting on a steep slope can and will likely experience landslides. The most mountainous states thus have the most landslides. States along the West Coast, in the Rocky Mountains, and in the Appalachians have the highest rates of landslides.

—USGS

HOW TO DETERMINE LANDSLIDE RISK

Landslides nearly always occur in identifiable hazard locations. You can know ahead of time if your house is vulnerable. If you are buying or building a new home, or if you want to know if your current home is at risk for landslides, here are some important factors to consider (from FEMA and the USGS):

- Do not build near steep slopes, close to mountain edges, streams and rivers near drainage ways, or natural erosion valleys.
- Get a geologic hazard assessment of your property.
- **Ask for information on landslides in your area and specific information on areas vulnerable to landslides, and request a professional referral for an appropriate hazard assessment of your property, and corrective measures you can take, if necessary.**
- Watch the hillsides around your home for any signs of progressive land movement, such as small landslides or debris flows or progressively tilting trees.

ARE THERE ANY WARNING SIGNS THAT A LANDSLIDE IS ABOUT TO OCCUR?

Sometimes landslides are preceded by the sound of branches breaking or boulders falling. La Conchita, California, resident Bill Harbison heard the loud *pop* sound before debris from the hillside came barreling down.

IN YOUR CAR

If you're driving along an embankment or a highway that abuts a bluff, here are signs to look for:

- If you are near a stream or channel, be alert for any sudden increase or decrease in water flow and for a change from clear to muddy waters.
- Watch the road for collapsed pavement, mud, fallen rocks, and other indications of possible debris flows.

> **DURING INTENSE STORMS**
>
> Stay alert and stay awake! **Many landslide fatalities occur when people are sleeping.** Listen to a radio for warnings of intense rainfall. Be aware that intense, short bursts of rain may be particularly dangerous, especially after longer periods of heavy rainfall and damp weather.

IN YOUR HOME

FEMA cites other landslide warning signs to watch out for:

- Changes occur in your landscape, fences, or retaining walls, utility poles slide in the land, or trees are progressively leaning.
- Doors or windows stick or jam for the first time.
- New cracks appear in the walls, exterior of the home, or the driveway.
- Bulging ground appears at the base of a slope.
- Water breaks through the ground surface in new locations.

The USGS says there are some steps you can take to make your home safer if you are living in a landslide-prone area:

- The most reliable protection is deflection walls, but most slides are large enough that it is not economically feasible to build physical protection.

LANDSLIDE SPEED

Debris flows (also referred to as mudslides, mudflows, or debris avalanches) generally occur during intense rainfall on water-saturated soil. They usually start on steep hillsides as soil slumps or slides that liquefy and **accelerate to speeds as great as 35 miles (56 kilometers) per hour**. Multiple debris flows that start high in canyons commonly funnel into channels. There, they merge, gain volume, and travel long distances from their source.

—USGS

- Slopes above structures can be stabilized by grading to lessen the slope steepness, by structural reinforcement, or by construction of buttresses or retaining walls to hold the slope in place. This can be rather expensive.
- Installing drainage systems either to prevent water from infiltrating into slopes or to allow water in the slope to escape can be very effective in stabilizing slopes.

WHAT TO DO IF YOU ARE CAUGHT IN A LANDSLIDE OR MUDSLIDE

According to the USGS's Dr. Jibson, "The 2005 La Conchita landslide moved very rapidly and was too fast for people to get away from, which is why several fatalities resulted." He offers more tips:

- If you are inside a structure and a rapid landslide hits, you should probably **take refuge in the interior of the home in the strongest part of the structure.**
- If you are outdoors, it is a matter of how fast the landslide is moving. Moving laterally out of the path of the landslide is probably more effective than trying to outrun it.

- Taking refuge behind buildings or trees might provide some protection. In general, **moving to higher ground** is a good rule of thumb.

LANDSLIDE SURVIVAL

"Accounts of those who have perished in landslides and debris flows often have a common theme: **the victim did not evacuate the house when they had the opportunity.** They may have assumed the house would provide some protection against landslides or debris flows. In practice, **most homes are not engineered to withstand the impact forces from many debris flows, and the home itself may become entrained with the debris flow**," says Dr. Jonathan Stock of the USGS.[6]

WHAT TO DO IF YOU SUSPECT IMMINENT LANDSLIDE DANGER

(FEMA)

- **Contact your local fire, police, or public works department.** Local officials are the best people to assess potential danger.
- **Inform affected neighbors.** Your neighbors may not be aware of potential hazards. Advising them of a potential threat may help save lives. Help neighbors who may need assistance to evacuate.
- **Evacuate.** Getting out of the path of a landslide or debris flow is your best protection.
- **Curl into a tight ball and protect your head if escape is not possible.**

Make sure your Family Disaster Plan is prepared and ready. It's a good idea to rehearse it as well. (Details on how to do this can be found in Chapter 13.)

WHAT TO DO AFTER A LANDSLIDE

Landslide debris flow does not come all at once. It arrives in bursts or pulses, so you may think it's over when it actually is not. This is why you do not want to go back to the original area where debris fell. Continual rainfall could trigger additional landslides. If the landslide was caused by a strong earthquake, the aftershocks that follow could cause more landslides. The USGS warns that this risk can occur minutes, days, weeks, or even months after the original earthquake. Here are its other tips:

- Stay away from the slide area. There may be danger of additional slides.
- Listen to local radio or television stations for the latest emergency information.
- Watch for flooding, which may occur after a landslide or debris flow.
- **Check for injured and trapped persons near the slide, without entering the direct slide area. Direct rescuers to their locations.**
- Help a neighbor who may require special assistance—infants, elderly people, and people with disabilities. Elderly people and people with disabilities may require additional assistance.
- Look for and report broken utility lines and damaged roadways and railways to appropriate authorities.
- Seek advice from a geotechnical expert for evaluating landslide hazards or designing corrective techniques to reduce landslide risk.

PROTECTING PETS IN LANDSLIDES

Bob Garcia is the supervising animal control officer with Sonoma County Animal Control in Santa Rosa, California. He's been working with animals for over 35 years. He has personal experience rescuing pets that were left behind and not evacuated during floods and landslides.

Garcia says pets often suffer injuries like lacerations, abrasions, sprains, fractures, and hyperthermia as a result of being left behind in a natural disaster.[7]

Garcia advises families that live in areas susceptible to landslides plan and prepare for how they would handle a quick evacuation. (Details on how to make your Family Disaster Plan that includes your pets can be found in Chapter 13. An emergency supply checklist for pets can be found in Chapter 15.)

It's important for pet owners to remember to take specific precautions for their animals. If you must leave a pet behind, be sure it has an escape route; never tie up an animal if floods are possible. **Some animals instinctively seek higher ground with the threat of a landslide, but others react defensively in fear.**

Pet owners should be prepared in the event of a disaster. This benefits the whole family, pets included.

CONCLUSION

Landslides occur in specific areas and repeatedly, so be sure to assess whether your house is in a precarious plot. Landslides or mudslides can occur during or after inclement weather as a result of dislodged ground from heavy rain. There are signs to watch out for that may indicate a landslide might occur. Loud rumbling of tree branches or a "popping" sound have been reported in the past by landslide witnesses just before the disaster. If ordered by local authorities, evacuate immediately. Have a plan in advance to know how you would care for your pets in an emergency. Make sure to contact local officials to find out when it's safe to return.

8
TSUNAMIS

Have you ever wondered what the bottom of the sea would look like if there were no water to cover it? It would be a floor of brown mud, dotted with stones and seashells. You might see small pools of water with flopping, brightly colored fish. Standing on the squishy, wet sand, you might be tempted to walk out toward the horizon, as the beach seems to go on forever into the sky.

On the morning of December 26, 2004, the Patong Bay pulled back to reveal this never-before-seen sight on the beach of Phuket, Thailand. The seafloor was now completely exposed.

Some locals climbed down from where they were perched on top of a six-foot retaining wall that divided the beach of Patong Bay from Thaweewong (Beach) Road. They were curious at the sight of the vast seafloor and got closer. They carried baskets to pick up the fish flopping around in small pools along the bare ocean floor. It was the day after Christmas, "Boxing Day," so the entire resort area was filled with tourists on their holiday. The white, sandy beach was packed with families enjoying the clear skies, before the heat of the day was at its most intense. Children played on the beach. The hundreds of people remained where they were, despite the sudden change in the landscape around them. Perhaps, they may have thought, this was an unusually low tide.

One of those Phuket vacationers was Richard Von Feldt, a 45-year-old Hewlett Packard executive. Richard grew up in a small town in Kansas, but he was now living the life of a world traveler. Based in Singapore at the time, Richard traditionally spent his year-end holiday at his favorite Thai hotel, the Baan Yin Dee. It was located on a hill to the south side of the beach. Richard liked it because it was close to the beach, but not in the center of the Phuket nightlife.

Earlier that morning, Richard had been on a plane from Singapore to Thailand, anticipating the relaxing beach vacation he'd planned. He was unaware that a 9.0-magnitude earthquake had occurred just before 8 a.m.

The quake happened in the Indian Ocean, 780 miles south-south-west of Bangkok, off the west coast of Sumatra. To get a better idea of where that is, see the map below. The star on the map indicates the location of the earthquake. The yellow dots point to the numerous aftershocks that followed.

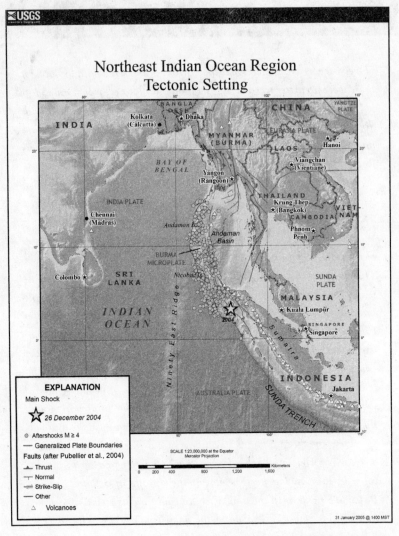

—*Courtesy the USGS*

*The quake was massive and originated in a **subduction zone** on the ocean floor. A subduction zone is an area where two tectonic plates come together and one overrides the other. According to the USGS,*

"The December 26, 2004 M=9.2 Sumatra-Andaman earthquake occurred along a tectonic subduction zone in which the India Plate, an oceanic plate, is being subducted beneath the Burma micro-plate, part of the larger Sunda plate."

"In other words," explains Dr. Kurt Frankel, an earthquake expert at Georgia Tech, *"it is a subduction zone setting where one plate is diving beneath another plate. This sort of plate tectonic setting is typically where the largest earthquakes occur."*[1]

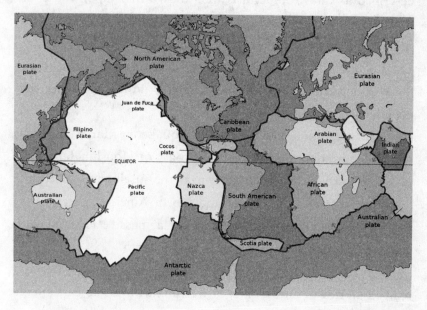

—*Courtesy the USGS*

THIS TYPE OF STRONG EARTHQUAKE IN A SUBDUCTION

zone can trigger tsunamis. This was evident in Chile in 2010 and Japan in 2011.

Despite the destruction that these earthquakes cause, the geological movement that causes them is very slow: "These two plates are converging at 7 meters per century."[2] (To put that in perspective, that's slower than your fingernails grow.) However, even that slow friction can eventually cause enough pressure in a subduction zone for a plate to pop up and displace the water in the ocean, triggering a tsunami. In Chile, the 2010 tsunami devastated several coastal towns in south-central Chile and damaged the port at Talcahuano, according to Sky News Online. They reported that "more than 210 people have been killed after a huge earthquake struck the Chilean coast, which triggered tsunami waves and sparked warnings for 53 countries."[3]

In Japan in 2011, the devastation that occurred can only be described as catastrophic. It is well known that Japan is a country prepared for earthquakes. Japanese homes and offices are designed with the region's highly seismically active location in mind. Their tsunami warning system is one of the most sophisticated in the world. **However, this earthquake was unlike any other they experienced in modern history.**

"The magnitude 9.0 Tohoku earthquake on March 11, 2011, which occurred near the northeast coast of Honshu, Japan, resulted from thrust faulting on or near the subduction zone plate boundary between the Pacific and North America plates," says the USGS. The Japan quake triggered a massive tsunami with waves that swept away everything in their path.

Witnesses captured dramatic video of the rushing water and relayed stories of entire villages washing away. **The earthquake and tsunami there resulted in one of the worst natural disasters in modern history.** It was not just a "double threat" but a triple one, as the earthquake damaged nuclear reactors in the area, and the Japanese people were forced to brace for a nuclear emergency in addition to the two deadly natural disasters they had just endured.

When the 9.0-magnitude earthquake struck off the coast of Honshu, it triggered a tsunami. The energy from the tsunami traveled across the Pacific.

From this one event in 2011, destruction and property loss were reported as far away as Hawaii and California.

One man on the U.S. West Coast lost his life. According to the Associated Press, "25-year-old Dustin Weber was swept out to sea at the mouth of the Klamath River by a tsunami surge generated thousands of miles away by the earthquake off Japan's coast. Rescuers were unable to reach him Friday and called off their efforts; he is presumed dead, though his body has not been found. His would be the first death of a person on the West Coast by a tsunami since 1964, when 11 people in nearby Crescent City died from the surge created by an earthquake in Alaska."[4]

In Japan, this disaster brought a truly staggering number of lives lost. Property damage is estimated to be in the hundreds of millions of dollars. This 2011 disaster illustrates the deadly power of tsunamis—just as the tsunami that hit Indonesia did just seven years earlier, on December 26, 2004.

On that morning, Richard Von Feldt did not know when he landed in Phuket that a major earthquake had just occurred in the Indian Ocean beneath him. He was in the air when the shaking was felt both on the ground in his departure city of Singapore and in his arrival city of Phuket.

Richard had taken the journey from the airport in a taxi to the hotel many times, and he settled back and popped open his laptop. Had he looked to his right, he'd have noticed the odd sight that caused his taxi driver to shout out words in Thai he didn't understand. The driver then came to an abrupt stop in the middle of the road. He got out of the car and gestured toward the Patong Bay. Normally the water was blue and sparkling. Even though Richard did not know what they were all shouting because it was not in English, he knew what had his driver and everyone else on the road so agitated. It was the fact that in the bay *there was no water*.

When Richard got out of the car, he found his driver outside staring in disbelief at the beach. The people he saw who were further down below the steep hill on the beach were not in their chairs with their eyes closed toward the sun. Many of them were gathered in groups, standing on the beach looking toward where the water was supposed to be. There were some children who seemed oblivious to the odd landscape around them. They were playing in the shallow puddles of what was once the water-covered seafloor.

Richard's thoughts and comprehension of the unusual sight were interrupted by the loud ring of his cab driver's cell phone. Richard watched the man answer the call. It was unclear who was on the other end, but the driver's face went from an expression of intense listening to confusion to sheer panic. According to Richard:

> He snapped that phone shut, and just started to run for the road—at the curb—to where his taxi was parked. I knew something was wrong. But mainly, I knew all of my luggage was in his car. And not sure of his motivation to run away, I was darn sure going to be near my luggage if that was where he was heading.
>
> He ran to his car and got in. I opened the door and also got in. I didn't even have my door shut, and he was already driving. He drove up onto the curb and started to weave along the sidewalk of people and tuk tuks (small cars) and motorcycles.
>
> He didn't say anything. But I knew he was making a run for something. We drove down toward the end of the beach. In the meantime, he pointed toward the water. In the far distance the horizon seemed to change color. Instead of a low, distant demarcation between sky blue and ocean blue, a new third color had entered the scene. The middle layer was two layers—one that was frothy white—and a second layer—a dark deep blue. It had somehow appeared out of nowhere. I didn't know what I was seeing. And each time I literally blinked my eyes to comprehend what I was seeing, the horizon came closer. Faster. Really fast.

Richard and his driver drove another 30 feet or so, climbing higher up the narrow, steep cement road. The taxi stopped again. They both got out and saw what was coming. Richard describes it:

As we stood there, it felt as if things were in slow motion. First, this large—wall is the best way to describe it—seemed to be traveling toward the beach at an inhuman rate of speed. Suddenly, people on the beach began to realize something wasn't right. And that this wave, which was confusing to judge by the eye, was not going to stop at the normal water's edge. Parents began screaming to their kids. Locals began screaming to each other. Everyone thought they needed to back up—way up on the sand. Yet even then, few realized that they needed to get completely off the beach—completely off the road—and as high as possible. Instead, everyone tried to just back up.

But as the water moved toward them, they backed up against a six-foot wall that separated the beach and the water. And unfortunately along that stretch of beach, there were only small sets of steps up that wall every half block or so.

The water rushed forward. Not in a wave. Everyone thinks it must have been a big wave. When we think of a wave, we think of something that is crashing and bubbling over. This was not a wave. This was a swell of water—a large push—as if a bathtub was just instantly filled up. On the morning of December 26, the bay filled like a bathtub. And it was a bad thing.

The water just kept coming. People thought it would stop, but it just kept coming. The water went to their knees. To their chest. Over their head. And over the six-foot wall. It crossed onto the sidewalk. Onto the road. And it kept coming.

We watched as the first push of water crested two to three feet over the road. Everyone was in disarray. Running. Screaming and confused.

If people could swim, they tried to stay on top of the water. But things started to float in the water. And if you grabbed anything that looked bigger than you, then it was dangerous, as you would be pushed up against other things that were just as large. And then the water seemed to hang motionless for a few seconds. It felt like someone turned off the water.

A moment after the water stopped, it started moving again. This time, though, it was heading in the opposite direction. Richard recalls:

All of that water started to suddenly pull back. It was like somewhere far out in the ocean, someone had pulled a drain plug. And the resulting suction of the water reversing direction was awful. Everything that was not tied down was being pulled back with the water. At that point, you could see people floating in the water, thinking that if they just swam to something, they would be safe.

But when the water started to recede, you didn't have a chance. All of the water that was high on the street rushed to the beach. And all of the water on the beach—over nine feet tall—also rushed back to the ocean, carrying with it people, motorcycles, and small debris from the road. And then it was gone. Within minutes, just as fast as that water came in, it returned back to the deep ocean. And once again, we saw the murky gray of the ocean floor in the bay. But this time, there were no fish. There were large wooden items. And large steel items. And bicycles. And small boats. And big boats. But no people. Those who were in the water and had not managed to hang on to a tree or something strong were also gone.

At that point, people were screaming. And many came running in every direction. Those in the middle of the ten-block area tried to run inland. Those close to us, at the end of the beach, came running toward the hill. People wanted to run up the hill to us. But down on the beach, they could not see what we could see. Out in the ocean, the water receded, and hung like a wall at sea. Just as fast as the water went out, it was on its way back.

But this time, it was larger yet. We could see how fast it was moving. And we could see that all of the people on the beach—on the road—in the storefronts and the hotel fronts were running in all directions.

The smart ones climbed floors to the roofs. A few ran and made it up our hill. But hundreds more were on the beach at ground level when the second wave came. And this time, it was 15 to 20 feet taller than the beach. The water swell rushed across the road, picking up everything in its way—including cars. Trucks.

It hit the front of any building lining the beach—and instantly wiped out every single lobby, every ground-floor hotel restaurant with people dining, and every person who was still sleeping in their hotel room. It filled every basement parking garage. It filled basement super-markets under the beachfront shops.

Richard says the water swirled and picked up everything in its path. He had moved farther up the hill, standing feet above where the water had stopped. He was ready to run to a higher perch if need be.

"Often, we simply held our hand out to people as they ran out of the water. We had to get them sat down, as many were in shock and delirious," he recalls.

More tsunami waves came over the next several hours, but they were smaller and smaller as time passed. Eventually Richard made it to his hotel, where he spent the night in a room with several others. There was no running water, little food, and no electricity. The next morning when daylight came, he took this picture on a street in Phuket.

—*Courtesy Rick Von Feldt*

The Indonesian tsunami in 2004 and the Japanese tsunami in 2011 unfortunately have shown that tsunamis can and do happen and that they are deadly. But what are tsunamis, and how do they happen?

The word *tsunami* comes from the Japanese words for "harbor" ("tsu") and "wave" ("nami"). The wave itself is caused by a sudden displacement of water under the sea. It can be caused by earthquakes, landslides, volcanic eruptions, or even meteorites. Earthquakes, though, are the most common cause. The energy travels from where the tsunami is initially triggered across the ocean.

Just above and in the immediate area around the earthquake in the ocean, the swell it may cause along the sea surface may be barely noticeable. **A ship that's near a tsunami wave may rise only minimally as it passes.** The problem is when that wave approaches the coast. It is moving at incredibly high speeds: 500 to 600 mph. The size and intensity of the tsunami wave are influenced by the depth and topography of the sea floor, known as bathymetry. Then, as it advances to the coastline, water has nowhere to go but up. As it rises, it pulls up the surrounding water (which is why the beach water recedes), and then **the wave comes barreling onto the coast as fast as a jetliner.**

Tsunami wave speed is controlled by water depth. Where the ocean is over 6,000 meters (3.7 miles) deep, unnoticed tsunami waves can travel at the speed of a commercial jet plane, over 800 km per hour (500 miles per hour). Tsunamis travel much slower in shallower coastal waters where their wave heights begin to increase dramatically.

—Pacific Tsunami Warning Center

Remember, tsunami waves are not like regular ocean waves for a variety of reasons. Tsunami waves are caused under the sea, typically by earthquakes. Regular ocean waves are driven by wind along the surface; they are also short-lived, as each one lasts for only 20 to 30 seconds. Compare that to tsunami waves, which can last for an hour or longer. This is why the ocean waves "break" at the beach and a tsunami wave comes in as a wall of water.

—University of Washington by the Department of Earth and Space Sciences

TIME OF ARRIVAL

 Locations nearest to the source of the tsunami will be impacted within minutes. Locations farther away may not see impacts for hours. Any coast can be impacted by a tsunami. Some areas are at a much greater risk than others due to the proximity to tsunami sources and seafloor."

—Jenifer Rhoades, National Weather Service
Tsunami program manager[5]

Proximity to the source proved deadly for those on the coast of Sendai, Japan, when the tsunami struck there in March 2011. One map that was widely featured in news accounts after the Japanese earthquake showed an approximate timeline of how long it would take for the tsunami to reach places like Hawaii, Chile, and California. It is based on a computer model that forecasts the potential arrival time. All day and night on March 11, 2011, television viewers around the world were watching "tsunami cams" pointed at the beaches for hours, waiting for any sign that the incoming wave had arrived.

You can see on the map that the further away you are from where the tsunami originated, the more time you have to prepare.

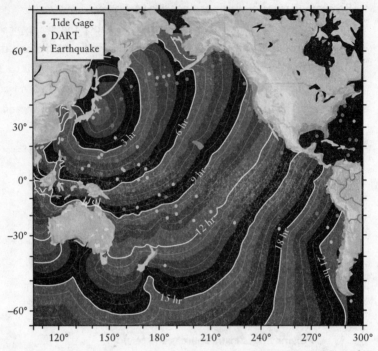

Tsunami Travel Times

Richard Von Feldt explained in his experience at Patong Bay in 2004 that the first tsunami wave that came in filled the beach with six feet of water. However, the second tsunami wave barreled in harder and brought the same beach a water height of about 15 feet. That second wave proved to be more deadly and destructive than the first. **"Tsunamis behave like a fast-rising flood tide, storm surge, or an advancing wall of water and strike with devastating force. They can move faster than you can run, and will continue for many hours after the first wave arrives. The first wave is often not the largest nor the most dangerous, and surges may arrive ten hours or more after the initial wave,"** says tsunami expert Jenifer Rhoades.[6]

Tsunamis are a series of waves, and because the second or third may be stronger than the first, unfortunately many people are misled into thinking the worst is over after that first wave hits.

Tsunamis are often no taller than normal wind
waves, but they are much more dangerous.

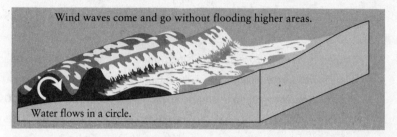

Wind waves come and go without flooding higher areas.

Water flows in a circle.

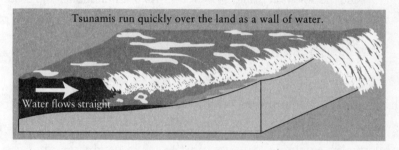

Tsunamis run quickly over the land as a wall of water.

Water flows straight

Even a tsunami that looks small can be dangerous!

Any time you feel a large earthquake or see a disturbance in the
ocean that might be a tsunami, head to high ground or inland.

Based on historical data, about 59 percent of the world's
tsunamis have occurred in the Pacific Ocean, 25 percent in the
Mediterranean Sea, 12 percent in the Atlantic Ocean, and 4 percent in the Indian Ocean, according to the USGS. There are fewer
tsunamis in the Atlantic and Indian Oceans because they have
fewer major subduction zones at the edges of plate boundaries,
unlike the Pacific, where the active "Ring of Fire" is located. (See
figure on page 189.)

The Mediterranean and Caribbean seas both have small subduction zones and histories of locally destructive tsunamis, according to the USGS. Only a few tsunamis have been generated in
the Atlantic and Indian Oceans.

Keep in mind that tsunamis can happen along any ocean coast. There is no season for them. They can happen anytime, day or night.

> " Tsunamis have been reported since ancient times. They have been documented extensively, especially in Japan and the Mediterranean areas. The first recorded tsunami occurred off the coast of Syria in 2000 BC. Since 1900 (the beginning of instrumentally located earthquakes), most tsunamis have been generated in Japan, Peru, Chile, New Guinea, and the Solomon Islands. Hawaii, because of its location in the center of the Pacific Basin, has experienced tsunamis generated in all parts of the Pacific.
>
> —USGS

Earthquakes are unpredictable, but there are warning centers around the world that monitor wave height in the event of an underwater earthquake. A rise in the wave height may indicate a tsunami has occurred. This work is done at data stations that use buoys and other instruments to measure sea surface levels. In the United States, the NOAA has implemented the DART (Deep-ocean Assessment and Reporting of Tsunami) program. The DART units are placed at sites with a history of generating destructive tsunamis.

> The United States has two tsunami warning centers: the Pacific Tsunami Warning Center in Hawaii and the West Coast and Alaska Tsunami Warning Center in Alaska. The Japan Meteorological Agency also has a well-developed tsunami warning system with multiple warning centers. **Following the 2004 Indian Ocean tsunami, other warning centers have been established in Indonesia, Australia, and New Zealand.** Efforts are currently under way to establish tsunami warning centers in the Caribbean and Mediterranean Seas.
>
> —Pacific Tsunami Warning Center

—*Courtesy NOAA*

In order to be prepared for tsunamis, it's important to know the watch and warning terms and what they mean. Here is a rundown of them from the **Pacific Tsunami Warning Center:**

> **Tsunami Information Bulletin/Statement:** A text product issued to inform people that an earthquake has occurred and to advise regarding its potential to generate a tsunami. **In most cases, a Tsunami Information Bulletin indicates there is no threat of a destructive tsunami, and is used to prevent unnecessary evacuations, as the earthquake may have been felt in coastal areas.**
> A Tsunami Information Bulletin may, in appropriate situations, caution about the possibility of a destructive local tsunami.

Tsunami Watch: The second-highest level of tsunami alert. Watches are issued by the TWCs based on seismic information without confirmation that a destructive tsunami is under way. It is issued as an advance alert to areas that could be affected by destructive tsunami waves.

Tsunami Warning: The highest level of tsunami alert. Warnings are issued because of the imminent threat of a tsunami from a large undersea earthquake or following confirmation that a potentially destructive tsunami is under way. Warnings advise that appropriate actions be taken in response to the tsunami threat. Such actions could include the evacuation of low-lying coastal areas and the movement of boats and ships out of harbors to deep water. Warnings are updated at least hourly, or as conditions warrant, to continue, expand, restrict, or end the warning.

If you are near the source of the tsunami caused by an earthquake, you can look to nature's warning signs. Tsunami expert Jenifer Rhoades says to be aware of what's happening around you. "Strong ground shaking, a loud roar from the ocean, the water receding from the shore unusually far and exposing the seafloor, or the water level rising rapidly are all nature's warning signs that a tsunami may be coming. If you observe any of these signs, immediately move to higher ground or inland. A tsunami may arrive within minutes," she says.

If you see any of these natural events occurring, Rhodes advises, "Do not wait for an official warning. The earthquake or changes in the water at the shore may be your only warning, and you may have only minutes to get to high ground. Stay away from low areas until told by officials that the danger has passed. Waves may impact the coast at irregular intervals for ten hours or longer.".

DO "TSUNAMI" AND "TIDAL WAVE"
MEAN THE SAME THING?

No, but unfortunately some people commonly refer to tsunamis as
"tidal waves," which is incorrect since tsunamis have nothing to do
with the tides. Tides result from the gravitational attraction of the
moon and the sun.

—Pacific Tsunami Warning Center

WHAT TO DO IF A TSUNAMI THREATENS YOUR AREA

The Mt. Ranier, Washington, chapter of the American Red Cross
has extensive tsunami preparation advice[7]:

Be familiar with the tsunami warning signs. A strong earthquake lasting 20 seconds or more near the coast may generate a
tsunami.

The most apparent sign that a tsunami is coming is a rapid rise
or fall in coastal waters.

Tsunamis most frequently come onshore as a surge of water
choked with debris. They are not V-shaped or rolling waves, and
are not "surfable."

Tsunamis may be locally generated or from a distant source.
In 1992, the Cape Mendocino, California, earthquake produced
a tsunami that reached Eureka in about 20 minutes, and Crescent
City in 50 minutes. Although this tsunami had a wave height of
about one foot and was not destructive, it illustrates how quickly
a wave can arrive at nearby coastal communities and how long
the danger can last.

PLAN FOR A TSUNAMI

For both local and distant events, you may learn a tsunami alert has
been issued by a Tsunami Warning Center through TV and radio

station broadcasts, **NOAA All Hazards weather radio**, or in some cases by announcements from emergency officials, aircraft, outdoor sirens, **Emergency Managers Weather Information Network (EMWIN),** or mobile devices. Immediately move away from beaches, harbors, or low-lying areas and follow instructions from emergency personnel.

—Jenifer Rhoades, NOAA

- **Develop a Family Disaster Plan.** Chapter 13 has details on how to prepare this plan for people and pets.

The Red Cross advises that tsunami-specific planning include the following:

- **Learn about tsunami risk in your community.** Contact your local emergency management office or American Red Cross chapter. Find out if your home, school, workplace, or other frequently visited locations are in tsunami hazard areas. Know the height of your street above sea level and the distance of your street from the coast or other high-risk waters. Evacuation orders may be based on these numbers.
- **If you are visiting an area at risk from tsunamis, check with the hotel, motel, or campground operators for tsunami evacuation information and how you would be warned.** It is important to know designated escape routes before a warning is issued.

This advice is echoed by NOAA tsunami expert Jenifer Rhoades: "When traveling to coastal locations, it is advisable to look for tsunami evacuation route signs and ask your local hotelier about tsunami safety for your location before a tsunami occurs. This will prepare you and your family to respond should a tsunami occur."

If you are in an area at risk for tsunamis (at home or elsewhere) do the following, according to the Red Cross:

- **Plan an evacuation route from your home, school, workplace, or any other place you'll be where tsunamis present a risk.** If possible, pick an area 100 feet above sea level or go up to two miles inland, away from the coastline. If you can't get this high or far, go as high as you can. Every foot inland or upward may make a difference. **You should be able to reach your safe location on foot within 15 minutes.** After a disaster, roads may become impassable or blocked. Be prepared to evacuate by foot if necessary. Footpaths normally lead uphill and inland, while many roads parallel coastlines. Follow posted tsunami evacuation routes; these will lead to safety. Local emergency management officials can help advise you as to the best route to safety and likely shelter locations.

- **Use a NOAA weather radio with a tone-alert feature to keep you informed of local watches and warnings.** The tone-alert feature will warn you of potential danger even if you are not currently listening to local radio or television stations.

Tulane University professor Stephen A. Nelson, associate professor and chair of the Department of Earth and Environmental Sciences, offers more advice below:

- "Never go down to the beach to watch for a tsunami! **When you can see the wave, you are too close to escape.** Tsunamis can move faster than a person can run!

- During a tsunami alert, your local emergency management office, police, fire, and other emergency organizations will issue instructions by radio or online. Give them your fullest cooperation.

- Homes and other buildings in low-lying coastal areas are not safe. Do **not** stay in such buildings if there is a tsunami warning.

- The upper floors of high, multistory, reinforced-concrete hotels can provide refuge if there is no time to quickly move inland or to higher ground.

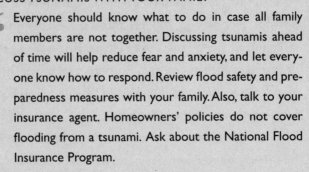

DISCUSS TSUNAMIS WITH YOUR FAMILY

" Everyone should know what to do in case all family members are not together. Discussing tsunamis ahead of time will help reduce fear and anxiety, and let everyone know how to respond. Review flood safety and preparedness measures with your family. Also, talk to your insurance agent. Homeowners' policies do not cover flooding from a tsunami. Ask about the National Flood Insurance Program.

—American Red Cross, Mt. Ranier, Washington, chapter

- **If you are on a boat or ship and there is time, move your vessel to deeper water** (at least 100 fathoms). If there is concurrent severe weather, it may be safer to leave the boat at the pier and move to higher ground.

- Damaging wave activity and unpredictable currents can affect harbor conditions for a period of time after the tsunami's initial impact. Be sure conditions are safe before you return your boat or ship to the harbor."[8]

WHAT TO DO AFTER A TSUNAMI OCCURS

From the Red Cross:

- **Use the telephone only for emergency calls.** Telephone lines are frequently overwhelmed in disaster situations. They need to be clear for emergency calls to get through.

- **Stay out of the building if waters remain around it.** Tsunami waters, like floodwaters, can undermine foundations, causing buildings to sink, floors to crack, or walls to collapse.

- **When reentering buildings or homes, use extreme caution.** Tsunami-driven floodwaters may have damaged buildings where you least expect it. Carefully watch every step you take.

- **Wear sturdy shoes.** The most common injury following a disaster is cut feet.
- **Use battery-powered lanterns or flashlights when examining buildings.** Battery-powered lighting is the safest and easiest, preventing a fire hazard for the user, occupants, and building.
- **Examine walls, floors, doors, staircases, and windows to make sure that the building is not in danger of collapsing.**
- **Inspect foundations for cracks or other damage.** Cracks and damage to a foundation can render a building uninhabitable.
- **Look for fire hazards.** There may be broken or leaking gas lines, flooded electrical circuits, or submerged furnaces or electrical appliances. Flammable or explosive materials may come from upstream. Fire is the most frequent hazard following floods.
- **Check for gas leaks.** If you smell gas or hear a blowing or hissing noise, open a window and quickly leave the building. Turn off the gas using the outside main valve if you can, and call the gas company from a neighbor's home. If you turn off the gas for any reason, it must be turned back on by a professional.
- **Look for electrical system damage.** If you see sparks or broken or frayed wires, or if you smell burning insulation, turn off the electricity at the main fuse box or circuit breaker. If you have to step into water to get to the fuse box or circuit breaker, call an electrician first for advice. Electrical equipment should be dried and checked before being returned to service.
- **Check for sewage and water line damage.** If you suspect sewage lines are damaged, avoid using the toilets and call a plumber. If water pipes are damaged, contact the water company and avoid using water from the tap. You can obtain safe water from undamaged water heaters or by melting ice cubes.
- **Use tap water if local health officials advise it is safe.**

- Watch out for animals, especially poisonous snakes, that may have come into buildings with the water.
- Use a stick to poke through debris. Tsunami floodwaters flush snakes and animals out of their homes.
- Watch for loose plaster, drywall, and ceilings that could fall.
- Take pictures of the damage, both of the building and its contents, for insurance claims.
- Open the windows and doors to help dry the building.
- Shovel mud while it is still moist to give walls and floors an opportunity to dry.
- Check food supplies. Any food that has come into contact with floodwaters may be contaminated and should be thrown out.

HOW TO PROTECT AND CARE FOR PETS IN A TSUNAMI

Dick Green, Ed.D., is the emergency relief manager of disasters for the International Fund for Animal Welfare (IFAW). He was on the scene in Southeast Asia to rescue animals in the 2004 tsunami. He was working in Sri Lanka on a vaccination program at the time.

In an exclusive interview for this book, he explains some of what he learned about animal behavior in that traumatic experience:

- Domestic animals at risk include dogs, cats, and backyard livestock. Unless they have a method of getting away, they will not likely survive the wave.
- Wildlife such as chimps, birds, and other wildlife that are tree-climbing may have a better chance of survival. And of course, birds will fare well on the initial impact but will suffer loss of habitat.
- Large wildlife like elephants and rhinos may not have the mobility to move as fast as cats. Interestingly, we saw very little evidence of these animals being lost in the Southeast Asia

tsunami, which leads to the question of whether they are able to perceive an oncoming wave.

- As animals lose their natural habitat or the human infrastructure that has provided food for them, they will resort to whatever methods they can to survive. **After the 2004 Pacific Rim tsunami, dog packs became more aggressive in finding food. They attacked people and created serious health issues.**[9]

As you would prepare for any extreme weather event or natural disaster, make sure you have your Emergency Pet Supply Kit on hand in the event of a tsunami. Try to make a smaller, portable version to take with you if you are traveling with your pet and staying in an area susceptible to a tsunami.

Note that if you ever have to evacuate, you should take your pet with you. If that is not possible, however, have a contingency plan for your pet in the case of an evacuation. Instructions on how to make a Family Disaster Plan for people and pets can be found in Chapter 13.

CONCLUSION

The recent tsunamis in 2004 and 2011 have brought the real threat of tsunamis to the forefront of people's minds. Most tsunamis are caused by earthquakes, but they can also be generated by volcanic eruptions, landslides, underwater explosions, and meteorites. If you hear on the news that an earthquake has occurred and you live in a coastal area, listen for the possibility of a tsunami advisory. This is also something to keep in mind when you travel. It's important to be familiar with the various types of tsunami advisories and especially to be alert for a **Tsunami Warning**. This is the most serious advisory and it means you must evacuate coastal areas immediately and seek higher ground. Plan ahead so you are familiar with an evacuation route if you need to get out quickly; this includes how you would get out of a hotel if you are on vacation. Though history shows us in some cases that animals instinctively know to move to higher ground before a tsunami hits, that does not always happen. Have evacuation plans ready for your pet as well.

9
SEVERE THUNDERSTORMS

The roar of the rain kept the others outside from hearing his screams. Something heavy and sharp had hit him hard on the left shoulder and back. The force of it had knocked down Todd Archer, a sportswriter for the Dallas Morning News, *and pinned him on his side.*[1] *Nothing was visible, because the thick, white tarp, which had once been the roof of the Dallas Cowboys' training facility, now completely covered his body. Todd shifted his weight about an inch, and if he looked to his left he could see a few inches off the ground. He could also feel the rush of the wind at the bottom of his legs. Were his feet exposed? Maybe someone could see him? He tried to move his right arm and leg. Nothing. Whatever was on top of him wouldn't budge. He was trapped. Todd called out again for help. "I'm under here! Help!!"*

Earlier that morning, before Todd Archer left his home in Flower Mound, Texas, he had practiced the same little ritual he did before work most days. He'd say to himself: "If I bring my umbrella, it will rain. If I don't bring my umbrella, it won't rain." On May 2, 2009, Todd left his umbrella at home.

It was one week after the NFL draft. Todd's assignment was to cover the first days of rookie mini-camp for the Dallas Cowboys at their training facility located in Irving, about ten miles outside Dallas.

Todd follows the Cowboys all the time, on and off the field. Wherever they go, he's there. Traveling the country for away games or charity events, he sometimes spends weeks away from his wife and two young daughters. His friends back home in Massachusetts thought the fact that he got paid to write about football was pretty cool—though they would have preferred he were covering the Patriots.

This time of year was particularly fun. It reminded Todd of the first day of school. The rookies, scouts, staff, and press were all together for the first time after the draft. He knew the Cowboys' press staff pretty well, along with many of the other sportswriters and television and radio reporters. The rookies, however, didn't know any of them, or each other. It was a time to catch up with his friends, but also to check out the new prospects. His assignment on this first Saturday in May 2009 was to write about the team's draft picks and how the rookies did in these early practices. He, along with dozens of other media members, traveled by car to Irving early that morning to cover the second full day of training camp.

On the drive into Irving, what was especially noticeable was the Cowboys' facility. It was the biggest building in the area filled with mostly strip malls. The white, tent-like structure juts over 80 feet into the sky and stretches for miles across an entire football field. It looked like a big white dome; that's why it's often referred to by the team as "the bubble."

There are two other uncovered football fields next to the tent-covered one, plus a building where the team had their lockers and the staff had offices and meeting space. Coaches and players prefer to use the exposed fields for their drills, as long as the weather was good. The morning started out fine; it was another mostly cloudy and muggy, late-spring Texas day, perfect for outside practice.

After a few hours of intense drills, they took a break for lunch. Work continued despite the break. Inside the main building, players, coaches, and scouts watched videos to analyze the morning's plays.

Soon it was time to head back out for more drills. By then, a little before 2 p.m., the skies looked more ominous. Gray clouds covered the sky, and it was a little breezier. The forecasts said there was a chance of rain in the afternoon. It was Texas and springtime. That's not unusual. Outside it looked as though the rain was coming sooner than later. It was decided to move afternoon practice into the "bubble."

Approximately 70 people, including players, reporters, photographers, coaches, scouts, and other Cowboys staff, moved inside through the glass doors of the large tent and metal structure.

Once inside, the team's videographers hoisted their cameras onto their shoulders and climbed the bleachers about halfway to the top of the tent. This way they could get aerial shots of the players. Todd and other reporters were near the end zone. The players moved to the center of the field.

In just a matter of minutes, the rain began. It didn't start lightly, either. From the inside, it seemed like the skies must have opened up, and it was now coming down hard and fast. The sound of the rain got louder. You could hear it pelting the top of the tent. Some people commented that they would all get soaked later when they headed back into the main building. Todd remembered that he'd left his umbrella at home.

After about 20 minutes, it was apparent that something strange was happening. The walls of the tarp were rippling—very slightly at first, then they puckered and shook violently.

Everyone looked up and saw an even more alarming sight. The heavy glass lights that hung 80 feet from the ceiling were swaying. They were large lights, suspended by wires. Normally they just hung straight down, and no one noticed them much. Now it looked like they could break off and fall at any moment.

A Cowboys' videographer was still up in the bleachers, perched 40 feet above everyone on the field. With one hand supporting the back of his camera and the other on the focus of the lens, he tilted back so his camera could capture the image of the large lights swinging like pendulums from the roof.

Todd and his colleagues saw the videographer still high up in his filming position. People began to shout: "He should get down. We need to get him down from there." Messages were yelled back and forth on two-way radios. The photographer started to maneuver himself and his equipment down the bleachers. Everyone was at ground level now. For just a few moments, the walls stopped shaking and the swinging lights slowed down a bit. Was it over?

Suddenly there was a flash of dark sky—where a wall had been seconds earlier. It seemed that the tarp had lifted to one side, exposing a daytime sky as black as night, illuminated only by a quick flash of lightning. The lights were back to swinging hard now, and everyone knew it was time to get out. Now!

Todd and a few others headed for the glass door in the right corner. He kept his cool, trying not to panic, not to press forward. They'd all get out okay, he thought. But something was wrong. The door wouldn't open. They pushed hard; it seemed glued shut. The force of the wind was blocking the door. Then it happened. The glass door popped and shattered. Broken glass was everywhere. Someone screamed, "Get out, get out now!"

Todd made it through the doorway, but he only took a few steps before something smashed into him from behind and knocked him straight to the ground. To make matters worse, Todd was shrouded with what once was the roof. The white tarp covered his body from *the knees up. He could see a few inches off the ground, but that was it. From what he could make out, there were lots of people running around, and he called out for help. But could anyone hear him?*

Todd wondered what the heck had just happened: "Was it a tornado? What's going on? What else could come down and hit me? Will I get trampled? Can anyone see me?"

They couldn't see all of him, but some of him . . . maybe. Todd's legs were exposed from the tarp from the knee down.

Back in 1996, he'd gotten a tattoo of a shamrock on his ankle. He'd done it in his 20s, just for fun. It was the tattoo that got Todd noticed and rescued.

Nick Eatman, a writer for DallasCowboys.com, saw someone's legs poking through the debris.[2] He later told Todd he had recognized Todd's shamrock tattoo on his ankle and knew it had to be him buried underneath. Nick tried to lift the metal piece of the door frame and the tarp that trapped his friend. He tried, along with another reporter from the team's website, Josh Ellis, but he couldn't move it.

Rookie players Brandon Williams, and DeAngelo Smith saw Nick trying to lift something. All four men pulled now. The debris was raised just enough for Todd to slide out.

In the meantime, Todd knew it was Nick trying to help him, because he recognized Nick's sneakers. He saw the cleats of two or three other players but found out who they were only hours later.

Todd managed to get to safety, though he suffered multiple injuries, including bruised ribs. Not everyone was so lucky. A coach as well as a Dallas Cowboys scouting assistant were severely injured in the collapse. When the news media reports showed how bad the damage was, most people agreed that it was amazing everyone survived.

WHAT EXACTLY HAPPENED AND WHY? WHEN TODD Archer was trapped, pinned down by heavy debris, he'd wondered if it was a tornado that caused the accident. It wasn't.

The National Weather Service of Dallas/Fort Worth determined that the collapse was caused by a *microburst*—an intense, localized downdraft of wind descending rapidly from a *severe thunderstorm*.

... roof collapse occurred, a *severe thunderstorm*
... was issued by the National Weather Service in Irving and
... region was under a Tornado Watch as well:

BULLETIN—EAS ACTIVATION REQUESTED

SEVERE THUNDERSTORM WARNING

NATIONAL WEATHER SERVICE FORT WORTH TX

3:06 PM CDT SAT MAY 2, 2009

THE NATIONAL WEATHER SERVICE IN FORT WORTH HAS ISSUED A

* SEVERE THUNDERSTORM WARNING FOR . . .

DALLAS COUNTY IN NORTH CENTRAL TEXAS . . .

* UNTIL 3:45 PM CDT

* AT 3:06 PM CDT . . . NATIONAL WEATHER SERVICE METEOROLOGISTS

DETECTED A LINE OF SEVERE THUNDERSTORMS CAPABLE OF PRODUCING

NICKEL SIZE HAIL . . . AND DAMAGING WINDS IN EXCESS OF 60 MPH.

THESE STORMS WERE LOCATED ALONG A LINE EXTENDING FROM SEVEN

MILES WEST OF COPPELL TO THREE MILES WEST OF CEDAR HILL STATE

PARK . . . AND MOVING EAST AT 40 MPH. PRECAUTIONARY/PREPAREDNESS

ACTIONS . . .

IF YOU ARE CAUGHT OUTSIDE . . . SEEK SHELTER IN A STRONG BUILD-

ING AND STAY INSIDE . . . AWAY FROM WINDOWS UNTIL THE STORM HAS

PASSED.

A TORNADO WATCH IS IN EFFECT FOR THE WARNING AREA. SEVERE

THUNDERSTORMS CAN PRODUCE TORNADOES WITH LITTLE OR NO

WARNING.

These severe thunderstorms can also produce microbursts
and tornadoes. While a microburst and a tornado both have
strong, destructive winds, they are not the same. Inside a tornado, the winds spiral upward from the bottom, sucking up dirt
and debris in their path. They do that from the narrowest part
of the funnel cloud at the bottom. By contrast, the winds in a

microburst are coming straight down from the cloud, sometimes pulling heavy rain right along with the wind. The powerful blast of cold air fans out and then curls back up when it hits the ground. The microburst is narrower at the top, and then as the winds are forced down rapidly the air and debris spread out wide at the bottom.

Downbursts in thunderstorms are also produced by rapidly sinking air in a severe thunderstorm. This term is used to describe a wider area affected than a microburst.

In the case of the May 2, 2009, roof collapse in Irving, Texas, the National Weather Service in Fort Worth reported that maximum winds near the ground from the microburst were estimated to be near 70 mph. It's possible, however, that they may have even been stronger. Wind speeds in these events often accelerate just before they hit the ground.

SEVERE THUNDERSTORMS

At any given moment, there are roughly 2,000 thunderstorms in progress around the world. It is estimated that there are 100,000 thunderstorms each year occurring in the United States. Warm, humid conditions combined with a lifting mechanism—like a cold front or even the heat of the day—can trigger the rumble of thunder and the burst of heavy rain. A thunderstorm is defined as a rain shower during which thunder exists. Since thunder comes from lightning, all thunderstorms have lightning. An average thunderstorm is 15 miles in diameter and lasts an average of 30 minutes. A thunderstorm is classified as "severe" when it contains one or more of the following: hail one inch or greater, winds gusting in excess of 50 knots (57.5 mph). About 10 percent of thunderstorms reach severe levels.

—National Severe Storms Laboratory

Since severe thunderstorms are so dangerous, it's important to be alert and aware if one is developing where you are. Here

> Severe thunderstorms contain multiple dangers that can threaten your safety and personal property in any part of the country and at any time of the year. Although these storms can be fairly small ... averaging around 10 to 15 miles in diameter ... and have an average lifetime of 20 to 30 minutes ... they can cause a tremendous amount of damage in this short time frame. Some supercell thunderstorms can have a lifespan of one hour or more.
>
> —NOAA

are thunderstorm advisories from the National Weather Service to familiarize yourself with:

Severe Thunderstorm Watch: This is issued by the National Weather Service when conditions are favorable for the development of severe thunderstorms in and close to the watch area. A severe thunderstorm by definition is a thunderstorm that produces hail that is one inch or larger in diameter and/or winds equal or exceeding 58 miles per hour, and/or a tornado. Severe thunderstorm watches are usually issued for four to eight hours. A watch is normally issued well in advance of the actual occurrence of severe weather. **During the watch, people should review severe thunderstorm safety rules and be prepared to move to a place of safety if threatening weather approaches.**

Severe Thunderstorm Warning: A severe thunderstorm warning is issued when either a severe thunderstorm is indicated by radar or a spotter reports a thunderstorm producing hail one inch or larger in diameter and/or winds equal or exceeding 58 miles per hour; therefore, people in the affected area should seek safe shelter immediately. If a tornado is present in a thunderstorm, a tornado warning will be issued instead of a severe thunderstorm warning. **Remember: severe thunderstorms can produce tornadoes with little or no advance warning.** Severe thunderstorm warnings are usually issued for one hour. They can be issued without a severe thunderstorm watch being already in effect.

A warning means the thunderstorms have already been reported and detected on radar and are imminent to occur. The watch means conditions are favorable for the weather event to happen. Watches often precede warnings (though they don't have to).

—*Courtesy NOAA*

LIGHTNING

Lightning is often a part of strong thunderstorms. It can be deadly.

In the United States, there are an estimated 25 million lightning flashes each year. During the past 30 years, lightning killed an average of 58 people per year. Yet because lightning usually claims only one or two victims at a time and it does not cause mass destruction of property, **it is underrated as a risk.**

—National Weather Service

While many of the locations where people were while they were struck by lightning go unreported, according to the Lightning Safety Institute, a number of the victims were in open fields or recreation areas, standing under trees, or engaging in outdoor sports on the water or land, and some were just using their telephone during a storm.[3]

MYTHS/FACTS ABOUT LIGHTNING

Myth: Rubber soles of shoes or rubber tires on a car will protect you from being injured by lightning.

Fact: Rubber provides no protection from lightning. However, the steel frame of a hard-topped vehicle provides increased protection from lightning (if you are not touching metal in the car).

Myth: People struck by lightning carry an electrical charge and should not be touched.

Fact: Lightning-strike victims carry no electrical charge and should be attended to immediately.

Myth: Lightning never strikes twice in the same place.

Fact: It can strike the same place over and over again. For example, the Empire State Building in New York City gets hit around 25 times a year!

Myth: Heat lightning occurs after very hot summer days and poses no threat.

Fact: What is referred to as "heat lightning" is actually lightning from a thunderstorm too far away for thunder to be heard. However, the storm may be moving in your direction.

—Source: NOAA/National Lightning Safety Institute

LIGHTNING STATISTICS

- Gender of victims: 84 percent male; 16 percent female. Note this does not mean women are less at risk for being struck by lightning.
- Months of most incidents: June, 21 percent; July, 30 percent; August, 22 percent.
- Days of week of most incidents: Sunday/Wednesday/Saturday
- Time of day of most incidents: 2 p.m. to 6 p.m.
- Number of victims per incident: One (91 percent), two or more (9 percent).
- Deaths by State, Top Five: Florida, Michigan, Texas, New York, Tennessee.

—Source: Lightning Safety Institute, based on a 1997 NOAA report for the years 1959–1994

Keep in mind that most people who are struck by lightning are not killed. Only around 10 percent of lightning strike incidents result in fatalities.

On January 10, 2011, 62-year-old Steve Chamberlin, a retired school administrator in Lake City, Florida, was struck by lightning on a golf course.[4] His group had come inside after playing a few holes because the horns had gone off, alerting them to severe weather. They went back out again after being given the all-clear. Not too soon after, Steve saw a distant flash in the sky.

"My friends and I decided to just finish the one shot to the green," Steve recalled. "I reached up for the club, and all of a sudden it was like a five-hundred-watt light bulb had gone off in my face. Then I was on the ground, and I felt burning on my lower back and down my legs. I couldn't feel my lower legs or my hands. I was really scared of what was to come. Was I going to be paralyzed? I was just lying there in the rain."

Steve's friends helped him to the golf cart, and once he was inside, he began to gain feeling again. His doctor checked him out and told him he'd likely be sore for a couple of days (which he was) but he would be okay.

"I am lucky to be alive," Steve said. "Someone up there is looking out for me."

Here is what you need to know to stay safe in severe weather, in a variety of places and situations.

OUTDOOR LOCATIONS WHERE YOU MIGHT BE VULNERABLE

Wide-open spaces can be dangerous because shelter can be far away. According to the Lightning Safety Institute, about 5 percent of lightning deaths in the United States annually occur on golf courses.

According to the Lightning Safety Institute, **"A good rule for everyone is: 'If you can see it (lightning), flee it; if you can hear it (thunder), clear it.'"** Its lightning safety tips include[5]:

- Suspend activities when lightning is 6 miles away.
- Avoid trees, poles, and other tall objects since they "attract" lightning.
- Avoid wet areas.

- Go to large, permanent buildings or get into a fully enclosed metal vehicle (car, van, or pickup truck).
- Do not touch any metal objects, including golf clubs, fences, electrical and maintenance machinery, and power lines.
- If sudden, close-in lightning does not permit immediate evacuation to a safer place, spread out from your group, squat, tuck your head, and cover your ears. Head for the safest place as soon as the immediate threat passes.

Parks and recreation areas are also dangerous places to be in a severe weather situation. The first advice is to monitor the local weather before and during your outing. Have a NOAA Weather Radio with you at all times. Do not rely on park employees to keep you abreast of the weather forecast; make sure you are alert and aware while you are outside. The same precautions for lightning safety on the golf course apply to any recreation area. However, the presence of numerous trees can create additional danger if you are in a park area.

At Yellowstone National Park, which spans Wyoming, Utah, and Idaho, visitors are advised to seek shelter immediately if caught in a thunderstorm[6]:

- If you are caught in a thunderstorm, do not use large boulders, trees, or other exposed large natural items for shelter; these objects are more likely to be struck by lightning due to their size.
- Never seek shelter in a tent, as the metal rods attract electricity. To reduce the risk of electrical shock, stay away from water.
- Although you should never lie flat on the ground during a thunderstorm, you should crouch down as low as possible.

Also keep in mind that if you are having a picnic, ducking under the picnic table won't protect you from lightning. Dugouts, sheds, and gazebos are also unsafe places. If you are with a group and there is no shelter, it's best to spread out and get into a crouched position. Remember—**do not lie flat on the ground.**

GOLFERS ARE ESPECIALLY VULNERABLE TO LIGHTNING

> " Some five percent of annual lightning deaths and injuries in the US occur on golf courses.
>
> —Richard Kithil, president and CEO, National Lightning Safety Institute

1. Golf is played in a wide-open area and golf clubs are metal—high conductors of electricity.
2. While on the course, players should stay aware of their surroundings and know how far they are from a safe place.
3. If on a golf course, do not touch any metal objects, including golf clubs, fences, electrical and maintenance machinery, and power lines.
4. The United States Golf Association (USGA) emphasizes that players in a competition have the right to stop play if they think lightning threatens them, even though the committee may not have authorized it specifically by signal (Rules 6–8 and 33–2d).

IF YOU ARE SWIMMING IN A POOL, A LAKE, THE OCEAN, OR ANY BODY OF WATER

Being in a body of water is one of the worst places to be during a thunderstorm. Water is a conductor of electricity. That means that even if you are in one end of the pool and lightning strikes the other end of the pool, you can still get injured.[7]

- Leave the water and seek shelter at the first signs of the storm.
- Swimming pools are connected to a much larger surface area via underground water pipes, gas lines, electric and telephone wiring, etc.

- Lightning strikes to the ground anywhere on this metallic network may induce shocks elsewhere.
- It is also unsafe to swim immediately after the storm passes.
- Pool activities should remain suspended until 30 minutes after the last thunder is heard.

IF YOU ARE IN A BOAT ON THE WATER

According to NOAA:

The vast majority of lightning injuries and deaths on boats occur on small boats with NO cabin. **It is crucial to listen to the weather when you are boating. If thunderstorms are forecast, don't go out.** If you are out on the water and skies are threatening, get back to land and find a safe building or safe vehicle.

Boats with cabins offer a safer, but not perfect, environment. Safety is increased further if the boat has a properly installed lightning protection system. If you are inside the cabin, stay away from metal and all electrical components. Stay off the radio unless it is an emergency.

If you are caught in a thunderstorm on a small boat, drop anchor and get as low as possible. Large boats with cabins, especially those with lightning protection systems properly installed, or metal marine vessels are relatively safe. Remember to stay inside the cabin and away from any metal surfaces.

If your boat does not have a cabin, crouch down in the lowest area. Do not dangle your legs or arms off the side.

IF YOU ARE IN YOUR CAR

If you are outside at a picnic or camping, it's advised to get to your car if another larger shelter is not nearby. But there are *safe* and *unsafe* vehicles. According to NOAA:

A safe vehicle is any fully enclosed metal topped vehicle such as a hard-topped car, minivan, bus, truck, etc. If you drive into a thunderstorm, slow down and use extra caution. If possible, pull off the road into a safe area. Do NOT leave the vehicle during a thunderstorm.

Unsafe vehicles include convertibles, golf carts, riding mowers, open cab construction equipment, and boats without cabins.

While inside a safe vehicle, do not use electronic devices such as radio communications during a thunderstorm. Lightning striking the vehicle, especially the antenna(s), could cause serious injury if you are talking on the radio or holding the microphone at the time of the flash. Emergency officials such as police officers, firefighters, and security officers should be extremely cautious using radio equipment when lightning is in the area.

HAIL

Another dangerous by-product of a severe thunderstorm is damaging hail. Hail is caused by small ice particles riding up and down the inside of a thunderstorm. The stronger the updraft, the longer the particle is suspended, and the larger the hailstone can grow.

> Large hailstones occasionally can be dangerous.... During a tremendous hailstorm in the Dallas, Texas, area in 1995, more than 100 people were injured by hail up to *four inches* in diameter. A hailstone the diameter of a golf ball (roughly one and three-quarter inches) can produce an injury, especially if it strikes a person's head. **A hailstone the diameter of a baseball (two and three-quarter inches) falls at a speed comparable to that of a pitched baseball ... on the order of 100 mph.** It's like being hit by a "bean ball" thrown by a Major League pitcher ... an injury from such a stone can be serious, or even fatal.[8]
>
> —Chuck Doswell, severe weather expert, certified consulting meteorologist

Large hail is especially dangerous for drivers, as it can break glass and even total a car. If you are driving and caught in a severe hailstorm, it's advisable to pull over until the storm passes. If it's a really bad storm, get down to the floor and cover your head and especially your eyes from breaking glass.

Even small hail can cause serious damage to crops and gardens. According to NOAA, hail causes one billion dollars of damage to crops and property each year.

Hail can occur in any state, at any time of day, in any season. You may hear on the radio or on TV that a storm is producing "pea-sized" or even "golf ball–sized" hail. The most common size for hail is one to one and a half inches. Here is a table to know how large or small the hail being described actually is:

Hail diameter	Compares to this object
1/4"	Pea or pencil eraser
3/4"	Penny
7/8"	Nickel
1"	Quarter
1 1/4"	Half dollar
1 1/2"	Walnut or ping pong ball
1 3/4"	Golf ball
2"	Hen egg
2 1/2"	Tennis ball
2 3/4"	Baseball
3"	Teacup
4"	Grapefruit
4 1/2"	Softball

WHAT TO DO IF YOU ARE AT HOME DURING A SEVERE THUNDERSTORM

The most important thing to do at home is to be aware. If severe weather threatens your area, it's important to stay informed.

Keep your NOAA Weather Radio turned on and keep listening for updates.

- Postpone your plans to go outside until the storm passes.
- At a time when the weather is calm, check your property for excess debris, like dead tree branches. Remove these items, as they can become hazardous in a storm.
- Remember that a severe thunderstorm can produce very damaging winds (as seen in the microburst example at the beginning of this chapter.)
- In case things become more dangerous, **always have your Emergency Disaster Supply Kit ready and your Family Disaster Plan organized and understood by everyone in the house.** (Details on how to build this kit and create a disaster plan for your family can be found in Chapters 14 and 13, respectively.)
- Keep away from windows and skylights. Hail and wind could shatter the glass.

> " If you're inside, you should stay off and away from all electrical appliances (including computers, washing machines, dishwashers, electronic games) and plumbing—including tubs, faucets, and, YES, toilets! During a thunderstorm do NOT unplug things—that should be done long before the storm is nearby if you're going to do it. Corded phones are dangerous during a storm. Do not use them. Cell phones are fine.[9]
>
> —Donna Franklin, NOAA's Lightning Safety Program Lead

According to NOAA, about 4 to 5 percent of people struck by lightning are struck while talking on a corded telephone.

- Before the storm, bring your car into the garage if possible to avoid hail damage.

- Wait until the storm is completely over before venturing outside to check for damage.

Note that it doesn't have to be stormy for lightning to strike.

BOLTS FROM THE BLUE

A lightning flash can travel horizontally many miles away from the thunderstorm and then strike the ground. These types of lightning flashes are called "Bolts from the Blue" because they seem to come out of a clear blue sky. While blue sky may exist overhead (or in part of the sky overhead) a thunderstorm is always located five to ten miles (and sometimes even further) away. Although these flashes are rare, they have been known to cause fatalities.

—Source: NOAA

WHAT TO DO IF YOU ARE WITH SOMEONE WHO HAS BEEN STRUCK BY LIGHTNING

According to the Red Cross, if you are with someone who has been struck by lightning, the obvious first thing to do is call for help. They advise[10]:

- Get someone to dial 911 or your local emergency medical services (EMS) number.
- The injured person has received an electrical shock and may be burned or have other injuries. People who have been struck by lightning do not retain an electrical charge and can be handled safely.
- Give first aid. If breathing has stopped, begin rescue breathing. If the heart has stopped beating, a trained person should give cardiopulmonary resuscitation (CPR). If the person has a pulse and is breathing, look and care for other possible injuries.

HOW TO CARE FOR YOUR PETS IN SEVERE THUNDERSTORMS

Planning ahead for how to care for your pet is crucial when it comes to severe weather. It's important to include pets in your **Family Disaster Plan** and have your **Emergency Disaster Pet Supply Kit** ready and accessible. Details on how to prepare a plan and build a kit are found in Chapters 13 and 15.

Have you ever noticed your pet acting oddly just before or during a thunderstorm? It's not uncommon, according to Joanne Yohannan, senior vice president of operations at the North Shore Animal League America: **"Animals can be afraid of unfamiliar loud noises, and thunderstorms certainly fall into that category. While some dogs may develop a fear of loud noises due to a traumatic experience that may be associated with the sound, others may develop fearful reactions for unknown reasons."**[11]

If you are wondering whether your particular breed is more susceptible than another to being afraid of severe weather, Yohannan says it's more about your specific pet than his or her breed. **"Any dog can be afraid of a thunderstorm. Every animal is an individual and may react differently to the storm—the challenge is desensitizing the dog to the sound."**

Yohannan says there are things pets owners can do before a storm and even during a storm that may help. Here are her tips:

- You can help your pet become familiar with the sounds of storms by routinely playing a thunderstorm CD at a low volume. Treats or even a calming massage at this time can help build a positive association.
- Always keep your pets indoors during a storm.
- Make sure your pet has an accessible "safe" place he can gravitate to during the storm, like a crate or a bed.
- Do not change your routine—act as you would during calm weather. Do not overcomfort your dog, as this may have

a reverse effect, making him even more nervous. This can reinforce his fears by sending him the message that the storm is something to be afraid of.

- Play soothing music.
- Distract your dog with play and toys.
- If you notice that your pet is showing severe signs of anxiety that are persisting, it is best to consult a veterinarian.

CONCLUSION

Severe thunderstorms can not only contain heavy rain, large hail, and strong winds, but they also can produce microbursts and tornadoes. While a tornado is a rotating column of air that sucks up everything in its path, a microburst is a fast-moving downdraft of wind. According to NOAA findings, it was a microburst, not a tornado, that damaged the Dallas Cowboys' training facility in 2009. A Severe Thunderstorm Warning means a thunderstorm has been spotted either by a weather observer or by radar, and you should take cover immediately. There really is no safe place to be outside during a thunderstorm. Lightning can strike on the golf course or even in the pool, so always seek sturdy shelter as soon as you hear the rumble of thunder. Follow the NOAA guidelines: WHEN THUNDER ROARS GO INDOORS. While many animals are fearful of thunderstorms, there are some things pet owners can do to make their pets feel more at ease in bad weather. Distractions like soothing music and toys might work to calm a pet. Always keep your animal indoors during thunderstorms; they are at risk of getting injured, just like humans, in a dangerous storm.

10

SNOW AND ICE STORMS

As 33-year-old Shaun Gholston drove to his job as a copy editor at a design firm in Rockville, Maryland, he encountered a cold, light rain. It was nothing more than a small nuisance. This was a typical winter morning on January 26, 2011. The ride from Shaun's home in northeast Washington, DC, to his office in suburban Maryland took about 45 minutes in the morning and just under an hour in the evening.[1]

Once Shaun arrived at work at his graphic design firm, snow was the topic of conversation. He'd heard a few of his coworkers saying that a storm was coming. No one mentioned a blizzard or anything really bad. It was more along the lines of some snow that might cause an inconvenience. As the day wore on, talk in the office switched from snow to when would be the best time to head home.

"I'll tell you what I did the last time it snowed," a young woman who works with Shaun told him. "I left after the rush, and I got home to DC even faster than I normally do!"

That made sense to Shaun. As he worked through lunch, he kept checking out his window, and it didn't look like much was happening. No reason to rush out.

His office cleared out at 3:30 but he stayed until 4:50, thinking that he'd beat the early rush of traffic. Shaun anticipated it would take maybe an additional hour to get home.

That was far from what happened next.

As Shaun left Rockville, the streets were covered in snow. He drove toward Montgomery Avenue. The snow began to get more intense, but traffic was moving. Shaun didn't see the weather as a huge problem . . . yet. That was until he approached the entrance ramp to the highway.

"The first red flag was that there was a line to get onto I–270," Shaun recalled.

This clearly was not a good sign. Once Shaun got onto that highway, cars barely crawled. The sky then burst open with heavy snow. It was sudden and treacherous. Snow was falling at a rate of two inches per hour.

"*Cars were literally sitting. We would move several feet, then sit for ten minutes. Move and sit, move and sit. Lanes became undefined,*" Shaun remembers.

Driving on I–270 to the I–495 (the Beltway) exchange usually took about 15 to 20 minutes. Three hours later Shaun had progressed only a few miles, and several tractor trailers had accidents on the Beltway. Dozens of cars were also stuck. Some drivers abandoned their vehicles, which made it even more challenging to remove them from the road. Fortunately, Shaun had filled up his car with gas the previous day.

Shaun was about two miles from the George Washington Parkway, when the man in the car behind him began honking his horn for him to move. The problem was, there was nowhere to go. Shaun was right behind an 18-wheel truck. Just minutes earlier, the truck driver had gotten out of his big rig to warn Shaun not to get any closer because his truck was stuck on a sheet of ice and could topple over at any moment. Shaun stayed where he was, but the honking did not stop.

Irritated, Shaun flung his car door open to a bitter blast of freezing air. "Dude! What the hell?!" he shouted at the car to his rear. "I obviously can't go anywhere!"

The other man got out of his car, and they were both yelling. The other driver was not happy. They stood face to face in the snow. Voices remained raised. Luckily, it didn't get physical. The other driver backed down and went back to sit in his car like everyone else. Traffic still wasn't moving.

It had been eight hours stuck at this point. To pass the time, Shaun called old friends on his cell phone. The sitting was becoming more uncomfortable. He got out of his car again to stretch.

Then something unexpected happened. There was a bright flash of light in the dark night sky. "It looked very strange," Shaun said. "I thought, 'This can't be lightning.' I was confused as to why the sky was lighting up but heard nothing. For a minute, I imagined what it would be like if the world was ending. That's what it started to look like, with all the stopped cars and flashing sky."

Even though Shaun couldn't hear the loud BOOM! of thunder, many others nearby did report it.

It was another two hours before traffic finally let up enough for Shaun to get off the Beltway and take an alternate route home.

When Shaun finally pulled into his driveway, it was was 4:09 a.m.—the day after he originally left work for his commute home. Shaun Gholston's normal 55-minute commute had taken 11 hours and 19 minutes.

THUNDER SNOW IS AN UNUSUAL WEATHER PHENOMENON.
It's the same type of thunder and lightning you see and hear in a
summer storm, but in the winter it's associated with snow, not
rain. While many warm weather days bring thunderstorms with
frequent thunder and lightning, most snow events don't bring
thunder snow. A combination of factors has to occur for it to
happen.

According to Chris Strong, a meteorologist for the Baltimore/
Washington forecast office of the National Weather Service,
"Lightning only needs a mix of liquid and ice in a cloud with
enough turbulence to create static electricity between the two. On
that day, temperatures were near freezing well up into the clouds,
and it was a very vigorous storm with enough convection heat to
create static electricity. While it's not common, thunder snow does
occur from time to time in strong winter storms."[2]

Shaun's commute was long not just because of the snow. It
was a combination of the weather and human nature. "Rain during the day negated any pretreatments that the department of
transportation used on the roads. . . . The winter event began
very rapidly with heavy sleet and snow . . . it came during rush
hour. Trees were knocked down at traffic choke points. Vehicles
were abandoned. Plows could not get around to do their work
when they were stuck in traffic with everyone else. No matter how
much warning you give, many people will not deviate from their
set plan based on a forecast. They will await the event, and then
act accordingly. In this case, if you waited for the snow it was too
late," says Chris Strong of the National Weather Service.

It may be easy to relate to this disastrous Washington, DC
snow storm if you live anywhere that's cold enough for snow.
Winter storms affect many states across the United States.
Unfortunately, every year people die in winter storms due to cold
weather exposure, vehicle accidents, and even heart attacks while
shoveling heavy snow.

> INJURIES DUE TO ICE AND SNOW
> - **About 70 percent result from vehicle accidents.**
> - **About 25 percent occur in people caught out in a storm.**
> - **Most casualties are males over the age of 40.**
>
> —Source: weather.gov

TYPES OF WINTER WEATHER/WINTER WEATHER ADVISORIES[3] (NOAA)

Sleet: Raindrops that freeze into ice pellets before reaching the ground. Sleet usually bounces when hitting a surface and does not stick to objects. However, it can accumulate like snow and cause a hazard to motorists.

Freezing Rain: Liquid rain when the temperature near the surface is below freezing. This causes it to freeze onto surfaces, such as trees, cars, and roads, forming a coating or glaze of ice. Even small accumulations of ice can cause a significant hazard.

Snow Squalls: Brief, intense snow showers accompanied by strong, gusty winds. Accumulation may be significant.

Blowing Snow: Wind-driven snow that reduces visibility and causes significant drifting. Blowing snow may be snow that is falling or loose snow on the ground picked up by the wind.

Blizzard: Winds of 35 mph or more with snow and blowing snow reducing visibility to less than a quarter mile for at least three hours.

Lake-effect Snow: Snow showers that are created when cold, dry air passes over a large warmer lake, such as one of the Great Lakes, and picks up moisture and heat.

Lake-effect Snow Squall: A local, intense, narrow band of moderate to heavy snow squall that can extend long distances inland. It may persist for many hours. It may also be accompanied by strong, gusty surface winds and possibly lightning. Accumulations can be 6 inches or more in 12 hours.

Wind Chill: Not the actual temperature, but rather how wind and cold feel on exposed skin. As the wind increases, heat is carried away from the body at an accelerated rate, driving down the body temperature.

Winter Storm Warning: Issued when hazardous winter weather in the form of heavy snow, heavy freezing rain, or heavy sleet is imminent or occurring. Winter Storm Warnings are usually issued 12 to 24 hours before the event is expected to begin.

Winter Storm Watch: Alerts the public to the possibility of a blizzard, heavy snow, heavy freezing rain, or heavy sleet. Winter Storm Watches are usually issued 12 to 48 hours before the beginning of a winter storm.

Winter Storm Outlook: Issued prior to a Winter Storm Watch. The Outlook is given when forecasters believe winter storm conditions are possible and are usually issued three to five days in advance of a winter storm.

Blizzard Warning: Issued for sustained or gusty winds of 35 mph or more, and falling or blowing snow creating visibilities at or below a quarter mile; these conditions should persist for at least three hours.

Lake-effect Snow Warning: Issued when heavy lake effect snow is imminent or occurring.

Lake-effect Snow Advisory: Issued when accumulation of lake effect snow will cause significant inconvenience.

Wind Chill Warning: Issued when wind chill temperatures are expected to be hazardous to life within several minutes of exposure.

Wind Chill Advisory: Issued when wind chill temperatures are expected to be a significant inconvenience to life with prolonged exposure, and, if caution is not exercised, could lead to hazardous exposure.

Winter Weather Advisories: Issued for accumulations of snow, freezing rain, freezing drizzle, and sleet that will cause significant inconveniences and, if caution is not exercised, could lead to life-threatening situations.

"   Winter events don't happen with minutes of lead time, like a tornado typically does. If people are connected, and able to receive warnings, there is usually enough advance notice for people to alter their daily plan and stay out of harm's way. It is those who try to "press it" and disregard warnings who most often find themselves in trouble.

—Chris Strong, warning coordinating meteorologist, NWS

HOW TO DRESS IN EXTREME COLD WEATHER

- Trapped air insulates. The best way to keep in heat is to wear layers. Outer garments should be tightly woven, water repellent, and hooded.

- Wear a hat. Half your body heat loss can be lost from your head.
- Cover your mouth to protect your lungs from extreme cold.
- **Mittens, snug at the wrist, are better than gloves.**

In order to be prepared for this potentially deadly season, here are some winter weather terms and advisories you should know:

DANGEROUS HEALTH EFFECTS FOR WHEN YOU FACE THE COLD OUTSIDE

Hypothermia: A condition brought on when the body temperature drops to less than 95°F, it can be fatal. For those who survive, there are likely to be lasting problems. Warning signs include uncontrollable shivering, memory loss, disorientation, incoherence, slurred speech, drowsiness, and apparent exhaustion.

Frostbite: Damage to exposed body tissue caused by extreme cold. A wind chill of −20°F will cause frostbite in 30 minutes. Frostbite causes a loss of feeling and a white or pale appearance in extremities, such as fingers, toes, ear lobes, or the tip of the nose. If symptoms are detected, get medical help immediately.[4]

—Source: NOAA

PREPARING FOR A WINTER STORM AT HOME

At home the biggest danger from winter storms is losing power and heat. In the cold weather months, it's especially important to make sure your Emergency Disaster Supply Kit is stocked and up-to-date. (Details on how to build this kit can be found in Chapter 14.)

Make sure your emergency heat sources, such as a fireplace or electric heater, are working properly. Also, have a fire extinguisher on hand, and don't leave it unattended. It is not advisable to heat your home using a stove or oven. For space heaters, make sure you follow the manufacturer's instructions to use them properly, and don't use them to dry wet clothing.

Frozen water pipes are a serious risk during very cold winter weather. When water freezes in a pipe it expands and can exert enough pressure to break it, causing a flood. The American Red Cross advises that you can prevent frozen pipes by taking some precautions in the cold weather season:

- Open cabinet doors to let warm air circulate around water pipes.
- Run the tap—running warm water through the pipe, even at a trickle, helps prevent pipes from freezing.
- Keep the thermostat set to a consistent temperature.

If you have no heat, NOAA advises:

- Close off unneeded rooms.
- Stuff towels or rags in cracks under doors.
- Cover windows at night.
- Eat and drink. Food provides the body with energy to produce its own heat.
- Wear layers of loose-fitting, lightweight, warm clothing.
- Remove layers to avoid overheating, perspiration, and subsequent chill.

Your NOAA Weather Radio should be turned on so you can be aware of advisories and changing conditions.

Chris Strong advises: "Check the forecast once a day. If there is threatening weather expected that day, check more frequently. Subscribe to text alerting services through local government to get additional information quickly. Sometimes extra minutes or hours count."

If you do have to evacuate due to extreme cold, have your Family Disaster Plan ready to enact immediately. (Details on how to make a family plan can be found in Chapter 13.)

PREPARE FOR A WINTER STORM IN YOUR CAR

As you read in the account of Shaun Gholston's harrowing 11-plus hours stuck in his car during a snowstorm, winter weather can cause nightmares for drivers. There are some steps you can take to be prepared.

This time of year, it's possible for your car to skid on ice or get stuck in the snow. Have an Emergency Disaster Kit in your car at all times. Details on how to prepare this kit can be found in Chapter 14.

Since you never really know when you will get stuck, keep your cell phone charged and your gas tank full. (A full tank of gas was really helpful in Shaun's and other Washington, DC, commuters' lengthy ordeals.)

Make sure your car is "winterized," meaning that you've had a mechanic or car expert properly check your automobile's fluids, wiper blades, tires, and other features to make sure your car is in good condition. Also, check that your car's tailpipe is not full of snow before you head out.

Ice is a huge hazard for winter drivers. At night, a light sheen on roads may look like rain when it is actually "black ice." Also, bridges and overpasses can ice before the rest of the roads do, as cold air wraps around them, cooling their surface temperature.

Driving in icy winter weather should be avoided. If you are in this situation, however, here are some tips from Bruce MacInnes, professional Formula Ford champion and professional driving teacher. Bruce has developed a method for driving on ice called the "skid control technique."[5]

"Driving in snow conditions requires a very light touch on all controls. In ice, drivers have almost no control if their speed is not extremely slow, compared to snowy conditions," MacInnes says. He adds that a big part of winter weather driving preparedness is tire choice. "Buy the best tires that you can afford. Good windshield wipers, and minus–30-degree washer fluid are also a safe bet."

Bruce advises you to keep plenty of distance between your car and the car in front of you. Also, "Do not turn with the brakes on. Avoid wheel spin when accelerating. Believe it or not, it's safer to get a running start approaching long hills. If you get stuck on a hill in a front-drive car, try safely turning around and backing up—the vertical load over the wheels will give you more traction. In rear-drive cars, sand bags in the trunk help traction."

Remaining calm and not panicking are also key. "The more relaxed you are, the wider your vision, and the better your balance. The best drivers are very calm." It's also advisable to keep someone informed of your route and where you are headed, just in case.

WHAT TO DO IF YOU ARE STUCK IN YOUR CAR IN A WINTER STORM

Mark Dornford, a staff sergeant in the United States Air Force, teaches cold weather survival to the Department of Defense personnel at one of the most frigid places on earth—arctic Alaska. If there is anyone who knows how to survive in cold weather, it's him.

Dornford says a winter survival kit for your car is crucial. "It should be tailored to the environment you find yourself in. If you live or are working in an area with cold temps and lots of snow, you should have a good sleeping bag, extra warm clothes, food, water (or a way to melt snow), metal container (to melt snow), candles, matches, winter hat, gloves, boots, flashlight, and some sort of signal light. I have all this stuff packed into a duffel back that I keep in the back of my car."[6] Full details on how to prepare a Winter Survival Kit for your car can be found in Chapter 14.

Here is Dornford's advice:

- Immediately ensure that the exterior exhaust on the car is clear of snow so you don't get carbon monoxide buildup in the car.
- Once you find yourself stuck in your car in a snowstorm, the best thing to do is stay put (unless you see a building nearby).
- Then turn the car off. Turn it on every once in a while to heat it up, and then turn it back off to conserve gas.
- Continue to check the tailpipe each time you turn the heater on.
- Although it sounds counterintuitive to exercise in the car, moving around will keep you warm. Kick your legs up or get in the backseat to do a set of pushups every 15 minutes.
- You also need to stay hydrated. If you don't have water in the car, you will need to melt snow. **The more hydrated you are, the warmer you will be, even if you're drinking freezing water from the melted ice.**
- Keep your seat belt on, and put on your hazard lights. Even if you're pulled over, people can still hit you.
- Another thing to think about is placing roadside flares near your car to alert other drivers/potential rescuers that you are there.

National Weather Service Meteorologist Chris Strong also advises motorists to "build flexibility into your plans: If there is threatening weather, avoid getting into a situation where you have to do things regardless of conditions."

WHAT TO DO IF YOU ARE STUCK
ON FOOT IN A WINTER STORM

At the Eielson Air Force Base in Moose Creek, Alaska, where Staff Sergeant Mark Dornford teaches military personnel how to survive if isolated in arctic elements, students spend several days a week outside in temperatures that range from −40°F to 20°F.

"One of the most extreme times I've had was teaching a class out here in Alaska in minus fifty-seven degrees (we don't have a temperature cutoff where we cancel training). Although we were outside all day and night, we didn't have any cold injuries. Another pretty crazy time was training people on the Greenland icecap when it was minus twenty with fifty-mile-per-hour sustained winds. Good times," recalls Staff Sergeant Dornford.

While most people will likely not find themselves in such extreme cold temperatures, Dornford offers some tips that anyone can use if they are stranded on foot in a snowstorm.

- **Dress properly for cold weather.** Wear layers and proper footwear. "A lack of preparation can lead to things like people being outside in a blizzard in tennis shoes or not having any gloves," Dornford says.
- **Make sure you stay hydrated.** "Dehydration is a very dangerous thing in a survival situation. Once you are dehydrated, you are much more likely to get frostbite and hypothermia. You will also not be thinking clearly and may start making poor decisions," according to Dornford.
- **Exercise to stay warm, but don't overdo it.** "If it is safe to walk outside, that is going to be a great way to stay warm. Just stay in sight of your car or camp. The last thing you need is to get lost. Just don't overwork to the point of sweating, because once

IS IT SAFE TO EAT SNOW IF YOU ARE THIRSTY?

" It is highly advised to melt snow before consuming it. Eating snow can significantly lower your core body temperature. However, if you are being very active, to the point of almost overheating, eating snow to stay hydrated isn't the end of the world. A better option would be to use that body heat from activity to melt the snow. If you have a sealable container, put it in between your layers of clothing near your core. You also need to check it periodically to make sure it doesn't spill. Once you begin to melt the snow, stir it with something often to melt it down quicker. It takes a lot of snow to procure a decent amount of drinking water. Snow to water is about a ten-to-one ratio. It is a very slow process. Still, it's better than nothing.

—Staff Sergeant Dornford, USAF

you stop moving you will be extremely cold. You also will need to change your socks if your feet start to get sweaty. Wet feet make you feel even colder. Calisthenics also are a good way to keep warm," Dornford says.

- **Keep a clear frame of mind.** "In a survival situation, it is vital to remain calm and avoid panicking. A good way to do this is to prioritize your needs. Look at what your most pressing needs are and start working on those first," according to Dornford.

HOW TO TAKE CARE OF YOUR PETS IN WINTER WEATHER

There is a popular belief that pets do fine outside in the winter because they have fur, but the North Shore Animal League America (NSALA), based in Port Washington, New York, says that's a misconception—they are as much at risk as we humans.

"They [pets] are susceptible to frost bite and hypothermia, which can be deadly. . . . When an animal comes in with hypothermia, they are usually shivering uncontrollably and their breathing becomes shallow. It is important that you warm the animal slowly—wrap the pet in blankets or even a heated blanket on a low setting if one is available. Put the pet in a heated car for the trip to the local veterinarian."

Joanne Yohannan of NSALA says smaller dogs are more vulnerable to the cold then larger ones. "Some dogs that are built for snow include Huskies, Malamutes, St. Bernards, Bernese Mountain Dogs, Great Pyrenees, Keeshonds, Collies, Labradors, and Newfoundlands, among others."[7]

Regardless of size, she says it's **important to keep your pet groomed.** "Believe it or not, knotted or matted hair doesn't insulate properly. Brush your dog's hair regularly, especially in the wintertime."

According to the ASPCA, more dogs are lost in winter than any other season because their sense of smell is weaker in the snow.[8] Make sure your pet has ID on at all times. Also, keep in mind that smaller dogs may have a tougher time walking in the snow than larger ones. And in case of an emergency, always carry your cell phone with you when you take your dog out for a walk.

If you are at home and lose power: "Dogs and cats are usually fine during the first few hours of a power outage. . . . Battery-operated space heaters can help keep pets warm. . . . Use extra caution if using candles or kerosene lamps, as curious cats can get too close and active pets can knock them over. If the outage forces you to evacuate, make sure to have an emergency bag with pet first-aid essentials packed and ready to go at a moment's notice. Being prepared ahead of time is the best bet," Yohannan advises.

(Details on what to have in your Emergency Pet Supply Kit can be found in Chapter 15.)

Here are more cold weather safety tips from the North Shore Animal League America:

- **All pets need to be inside in the winter.** Never leave your pet outside in the cold, even in a doghouse. When the temperature drops, your pet can freeze to death. If you notice a pet being locked outside in the winter, be sure to report it to your local animal control facility.

- **Antifreeze and rock salt are poisonous to your pets.** Make sure to keep these and other harmful chemicals out of your pet's reach or path. You may want to look into finding more eco- and pet-friendly products.

- **Pets should not be left in the car.** Most people know not to leave their pets in a car in the summer, but the same goes for the winter. A car interior can get as cold as an icebox and a pet can easily freeze.

- **Be sure to wipe your dog's feet (and stomach in small dogs) after a winter walk.** Rock salt or other ice-melting chemicals can cling to your pet's fur, and he can ingest these poisonous chemicals when cleaning himself.

- **Check your dog's paw pads for ice balls.** If your dog is lifting his feet a lot or seems to be walking strangely, his feet are probably too cold or ice may be forming, which can cause frostbite.

- **Do not let your pet eat snow.** Harmful chemicals or objects could be hidden in the snow, and pets could also develop hypothermia from eating the cold substance.

- **Short-coated dogs are especially vulnerable to the cold** and shouldn't be outside unattended or for too long.

- **Keep an eye on your pet's water dish to make sure the water did not freeze.**

CONCLUSION

Winter brings the beauty of snow-covered landscapes, but it can also bring danger to the people who live in cold and snowy climates. Exposure to cold temperatures for a prolonged period of time can result in frostbite or hypothermia. Shoveling snow can also pose a health hazard. Pay close attention to winter weather advisories and plan your winter travel in accordance with the forecast. Power outages can last for days, even weeks, after a winter storm. Prepare for the worst before winter starts. That includes a plan for your pets. According to the ASPCA, more dogs are lost in winter than any other season.

11
EARTHQUAKES

Emmanuel Buteau was sure he was dead.

"I thought maybe I was a ghost. It was black all around. I was confused. Was I alive or not? . . . I prayed to God."[1]

For 11 days, the 21-year-old Haitian man was trapped in the rubble of his collapsed Port-au-Prince home. He had no food or water. Since the devastating 7.0-magnitude earthquake's epicenter was only ten miles West of Haiti's densely populated capital city on January 12, 2010, Emmanuel drifted in and out of consciousness.[2]

Occasionally he heard voices around him speaking Creole, French, English, and Hebrew.[3] Those were the languages of the international rescue workers. Emmanuel worked as a tailor and lived with his mother, Marie Yolene Bois de Fer, in one of the poorest section of the city.[4] He said he was upstairs in his family's three-story home about to take a bath that afternoon in January.[5]

"Then I heard a huge noise. The house was dancing. It was shaking and then it all came falling down," he said.[6] "I was knocked to the ground. It was completely dark all around."[7]

The building collapsed with Emmanuel inside it. His mother, Marie, who was not at home at the time, never gave up hope of finding her son. At her insistence, an Israeli rescue team worked through the debris for hours until they saw Emmanuel's hand.[8] It was moving.

When Emmanuel heard the sound of his mother's voice, he knew then that he wasn't a ghost. "Mama, mama, mama," he shouted. "I'm alive!"[9]

THE RED CROSS ESTIMATES THAT THREE MILLION PEO-ple, one-third of the country's entire population, were directly

affected by the earthquake in Haiti. The tragedy was especially catastrophic due to the proximity of the epicenter (only ten miles) west
of the heavily populated capital, Port-au-Prince, combined with the
shoddy building construction in this poverty-stricken nation.

Dr. Kurt Frankel, an earthquake expert at the Georgia Institute
of Technology, says Haiti is not the only dangerous place to be in
an earthquake.

"Places like Istanbul, Pakistan, India, China, and Mexico
City are not up to the standard building codes we see in more
earthquake-prepared places like Tokyo or Los Angeles," he says.[10]
"Construction codes are very strict, for example, for everything
from bridges to apartment buildings, in earthquake-prone areas
like California."

The United States Geological Survey, an agency that monitors earthquakes, estimates that **several million earthquakes occur
around the world each year.** Many go undetected because they
are so minor on the magnitude scale that you don't feel them.
However, **a damaging earthquake can strike certain parts of the
world at any time without warning.**

Even though earthquakes can't be predicted, there is a lot
you can do to prepare yourself and your family—including your
pets—for when this natural disaster occurs.

WHAT CAUSES AN EARTHQUAKE

An earthquake is caused by the constant movement of tectonic
plates, and are most common where these plates meet, or "faults."
A sudden slip along a fault can cause a catastrophic earthquake.
Tectonic plates are slowly moving, but they get stuck at their
edges due to friction, which can cause an earthquake. The energy released travels in waves through the earth's crust and causes
shaking at the surface where we live.

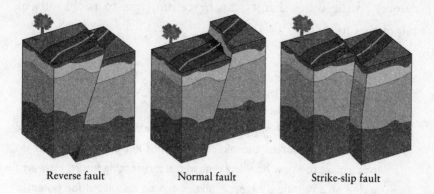

Reverse fault   Normal fault   Strike-slip fault

Above: Three basic fault types. *Strike-slip faults,* such as the San Andreas fault in California, are vertical (or nearly vertical) fractures where the blocks have mostly moved horizontally. *Normal faults* occur when the blocks have mostly shifted vertically. If the rock mass above an inclined fault moves down, the fault is termed normal, whereas if the rock above the fault moves up, the fault is termed *reverse.* (Source: soundwaves.usgs.gov)

PLATE TECTONICS

For hundreds of millions of years, the forces of continental drift have reshaped the earth. Continental drift is based on the idea that the continents bumped into and slid over and under each other and at some later time broke apart. Today, we know that the earth's crust is on the move, and we call this theory *plate tectonics.* Plates move mere inches annually, carrying the continents with them as they drift about.[11]

WHAT IS A FAULT?

Earthquakes occur on faults. A fault is a thin zone of crushed rock separating blocks of the earth's crust. When an earthquake occurs on one of these faults, the rock on one side of the fault slips with respect to the other. Faults can be centimeters to thousands of kilometers long.[12]

CAN EARTHQUAKES BE PREDICTED?

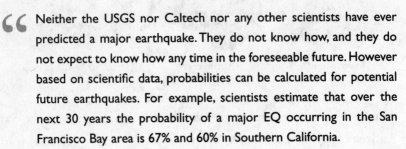

Neither the USGS nor Caltech nor any other scientists have ever predicted a major earthquake. They do not know how, and they do not expect to know how any time in the foreseeable future. However based on scientific data, probabilities can be calculated for potential future earthquakes. For example, scientists estimate that over the next 30 years the probability of a major EQ occurring in the San Francisco Bay area is 67% and 60% in Southern California.

—USGS

MAGNITUDE SCALE

When earthquakes are reported on the news, a magnitude scale number is used to classify the intensity level. The Moment Magnitude Scale is what's used today instead of the Richter Scale. Earthquake expert Dr. Kurt Frankel of Georgia Tech explains: "We no longer use the Richter scale because it is only good for local events (i.e., those that occur close to the actual earthquake). Once you get past a magnitude–6 earthquake the scale does not do a good job of recording moderate to large events. We now mainly use the Moment Magnitude Scale, which measures the amount of energy released during an earthquake. This is a much more accurate measure of earthquake size."

Earthquake Magnitude Classes
(USGS)

Great:	M 8+
Major:	M 7 to 7.9
Strong:	M 6 to 6.9
Moderate:	M 5 to 5.9

Light: M 4 to 4.9
Minor: M 3 to 3.9
Micro: M

WHERE DO EARTHQUAKES OCCUR?

Earthquakes occur at any place at any time. About 90 percent of the world's earthquakes occur along the rim of the Pacific Ocean; this zone is called "The Ring of Fire." It extends from Chile northward along the South American coast through Central America, Mexico, the West Coast of the United States, and the southern part of Alaska, through the Aleutian Islands to Japan, the Philippine Islands, New Guinea, the island groups of the Southwest Pacific, and to New Zealand.

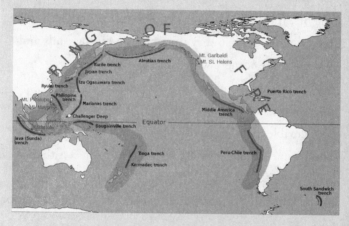

The "Ring of Fire" is also called the Circum-Pacific belt. About 5 to 6 percent of earthquakes occur from the Mediterranean region eastward through Turkey, Iran, and northern India.

—USGS

HOW TO PREPARE FOR AN EARTHQUAKE
(Daretoprepare.org, Shakeout.org, SCEC.org, USGS)

- Bolt and brace water heaters, refrigerators, and gas appliances to wall studs.

- Bolt bookcases, china cabinets, and other tall furniture to wall studs.

Secure your space: In the 1994 Earthquake in Northridge, California, a UCLA study cites that of the more than 9,000 people who were injured, 55 percent of the injuries were caused by falling objects in their homes like pictures, lamps, books, mirrors, and decorative items.

—Source: Daretoprepare.org

- Install strong latches on cabinets so glass and other items can't fall out.
- Gas appliances should have flexible connectors to reduce the risk of fire.
- Mirrors, framed pictures, and other objects should be hung from closed hooks so they can't bounce off the walls and secured at the corners with earthquake putty.
- Don't hang heavy items over beds or couches.
- Televisions, stereos, computers, and microwaves can be secured with flexible nylon straps and buckles.
- Check your garage or utility room for any objects that can fall, spill, or leak.
- Move flammable or hazardous materials to lower shelves or the floor.
- Learn about the nearest exits and evacuation plans for any place you frequent.
- Pick a safe place in each room of your home or office that you can move to quickly, such as under a piece of furniture, away from windows or anything that could fall on you.
- Prepare your Emergency Supply Kit and your Family Disaster Plan (complete details on how to do this can be found in chapter 13 of this book).

WHAT TO DO WHEN THE SHAKING STARTS

If you are inside during an earthquake:

- DROP, COVER, and HOLD ON.
- Move only a few steps to a nearby safe place.
- Take cover under a piece of heavy furniture or stand against an inside wall.
- Stay away from windows.
- If you are in bed, hold on and protect your head with a pillow.
- Many people think that doorways are the best place to be during an earthquake, but they are actually no stronger than any other part of the structure.
- Stay indoors until the shaking stops. Don't try to run outdoors.

On March 10, 1933, Southern California experienced a deadly seismic disaster when a magnitude–6.4 earthquake struck the Long Beach area. The earthquake killed over 120 people and caused property losses estimated at $50 million (1933) dollars. More than two-thirds of the 120 deaths occurred when people ran outside and were struck by falling bricks, cornices, parapets, and building ornaments.

—The Southern California Earthquake Center (SCEC)

- When you do exit, use the stairs, not an elevator.
- Be prepared for potential aftershocks after the initial quake.

If you are outside during an earthquake:

"An ideal spot to be during an earthquake," says Dr. Kurt Frankel of Georgia Tech, "is an open field, where nothing can fall on you."

- If you are in a crowd, try to move away into an open area.
- Stay away from buildings, trees, and power lines.
- Drop to the ground and cover your head and neck with your arms until the shaking stops.
- If you are in a mountainous area or near unstable slopes or cliffs, be alert for falling rocks or debris. Landslides are often triggered by earthquakes.

If you are in your car during an earthquake:

According to Dr. Frankel, "Unless it's a major quake, you probably won't feel anything while driving."

- If you feel the shaking slow down and drive to a clear place.
- Turn on emergency flashers.
- When you come to a stop, turn off the ignition and set the parking brake.
- Stay inside the car until the shaking stops.
- Avoid bridges, ramps, and overpasses. Do not drive on them or under them during an earthquake if possible.

HOW TO PROTECT YOUR PETS DURING AN EARTHQUAKE

Scientists say earthquakes can't be predicted. Others may tell you that animals, judging by their reportedly sometimes odd behavior, know when one is coming. Earthquake experts argue that this may be because animals are more sensitive to the first tremors, called the first tremors that occur before the violent shaking starts.

While your pet may not be able to warn you of an impending earthquake, here is what you need to know to keep him or her safe[13]:

- If you have a crate or a dog bed for a pet, do not place it below windows or any objects that can fall on it.
- Tighten the latch on your birdcage so that the door cannot be shaken open easily.
- Secure pet cages and aquariums on low stands or tables, as they may move or break during a disaster.

" Train your pet to enter his/her carrier or crate at your command. Try putting your pet's favorite treat in his/her carrier and sounding a bell at the same time. Repeat this process every day, until your pet comes running at the sound of the bell. Continue this routine often enough to keep it fresh in your pet's mind. This training will be extremely helpful when locating a frightened animal.

—Mary McCarthy, pet sitter, athomepetsittingsf.com

ODD ANIMAL BEHAVIOR BEFORE AN EARTHQUAKE

- In 373 BC, historians recorded that animals, including rats, snakes, and weasels, deserted the Greek city of Helike in droves just days before a quake hit.[14]
- In the winter of 1975, Chinese officials ordered the evacuation of the city of Haicheng in northeastern Liaoning Province the day before a 7.3-magnitude earthquake, based on reports of unusual animal behavior and changes in ground water levels.[15]
- On August 23, 2011, a 5.8-magnitude earthquake occurred nine miles south of Mineral, Virginia, at 1:51 pm. Just a few minutes earlier, some strange animal behavior was reported the National Zoo in Washington, DC:
  - About five to ten seconds before the quake, many of the apes, including Kyle (an orangutan) and Kojo (a Western lowland gorilla), abandoned their food and climbed to the top of the tree-like structure in the exhibit.
  - About three seconds before the quake, another gorilla, Mandara, let out a shriek and collected her baby, Kibibi, and moved to the top of the tree structure as well.
  - The black-and-rufous giant elephant shrew hid in his habitat and refused to come out for afternoon feeding.
  - The zoo has a flock of 64 flamingos. Just before the quake, the birds rushed about and grouped themselves together. They remained huddled during the quake.[16]

- While you may want to hold your pet during the violent shaking of an earthquake, be careful because the animal may want to hide and may bite or scratch you to protect itself.
- Says Mary McCarthy: "Watch the body expressions and sounds your pet makes to warn you. Even your own pet can be

aggressive when in pain or frightened. A cat might crouch with ears flattened to its head. A dog might snarl and its hair might rise on its back/hind. Yes, even your favorite family pet can behave this way. Quick, jerky movements, or loud voices also can scare your pet."

- Keep your dogs in a fenced yard so you can keep an eye on them.
- If you need to evacuate with your pet, have your Emergency Pet Supply Kit ready to go. (Details on how to prepare this kit can be found in the emergency checklist for pets in Chapter 15.)
- Remember that many shelters don't accept animals, so have alternate plans in place, like a list of kennels or dog-friendly motels, just in case.
- Make sure your pet is always wearing an ID tag and, if possible, an internal ID, such as a microchip.
- If you can't find your pet and need to evacuate immediately, prepare an emergency pen for pets at home that includes a three-day supply of dry food and a large container of water.
- Low-fat, dry food is best because it deteriorates more slowly.
- Leave a note for rescuers who may come to your home or find your pet as to where you are and how to contact you.

CONCLUSION

While it's impossible to predict where and when an earthquake will strike, certain parts of the world are more prone to them than others. Know your area's history and propensity for quakes and follow the instructions on how to best quake-proof your home. If you live in such a region, secure your household items so they are less likely to cause injury if violent shaking occurs. Having a disaster plan will prepare you and help care for your pets in times of danger when panic might otherwise take over.

12
HOW TO USE SOCIAL MEDIA IN NATURAL DISASTERS

On May 22, 2011. A large portion of Joplin Missouri was devastated by an EF–5 (winds greater than 200 mph) *tornado killing over* 150 people and injuring over a thousand others *in the Joplin, Missouri, area. The tornado was ¾ of a mile wide, and had a long track on the ground of six miles. According to NOAA, the Joplin tornado is the deadliest since modern recordkeeping began in 1950 and is ranked eighth among the deadliest tornadoes in U.S. history.*

For those who survived this tragedy, all that was left were destroyed homes and missing belongings. While it wasn't possible to reclaim all that was lost, there was a way to search for the smaller possessions of sentimental value. Debris traveled many miles in this epic disaster. Family photos and other belongings had scattered in the storm, showing up on neighbors' property. Not only were Joplin residents looking for these items; they also wanted to reunite what they found with the rightful owners.

That's why the Facebook page Joplin Tornado Debris: Lost and Found[1] *was created.*

Even months after the catastrophic tornado decimated the city of Joplin, Missouri, on May 22, 2011, posts on this social media site are uploaded daily:

> Desperately searching for this lost quilt from a home at 4400 Grace Lane. Love Eagle quilt made in 1930. Cheddar gold/ivory. 21 Eagle blocks/21 airplane blocks. This is a treasured item. Any help would be so appreciated.

> I found a Star Trek wall clock. It's a little weathered but still one of those things you don't really want to lose. If it belongs to anyone, let me know. I'm getting the parts to fix the mechanism this week.

So glad someone is doing this. Small things can mean so much when you have lost everything.

Comments follow posts with more details and updates are listed when a match is made.

In the wake of the devastating earthquake and deadly tsunami that struck Japan on March 11, 2011, another social media tool, Twitter, served as a primary way between survivors and their friends and family to communicate.

Twitter, which allows users to post very short messages—no longer than 140 characters—became popular crucial resource for people trying to find out news and information. People in Japan used it to post news about how serious the situation was where they were, along with uploads of mobile videos they had recorded.

Within an hour, more than 1,200 tweets a minute were coming from Tokyo. By the end of Friday, American time, a total of 246,075 Twitter posts using the term "earthquake" had been posted. [2]

In the hours that followed the earthquake in Haiti on January 12, 2010, Facebook employees, including marketing director Randi Zuckerberg and Andrew Noyes, manager of public policy and communications, quickly set up the Global Disaster Relief Page *(facebook.com/DisasterRelief).* [3] *"The days and weeks after the quake underscored the internet's critical role in connecting the world's population in times of tragedy. The* Global Disaster Relief *page on Facebook became a destination for millions of people who wanted to educate themselves and find out how to help not only Haiti, but wherever disaster and misfortune may strike," said Noyes.*

THERE IS SO MUCH SOCIAL MEDIA TECHNOLOGY AVAIL-able that it can be overwhelming to determine what to use and how to use it in the aftermath of a natural disaster. That's why Sree Sreenivasan, a social media expert who's been teaching digital media classes at Columbia University since 1994, says the most important thing you can do with social media is to become familiar and comfortable with the medium *before* an extreme weather event occurs.[4]

> Understand the techniques, the language, and the etiquette in advance. You want to be able to understand how to share news during earthquakes, tsunamis, tornadoes, and hurricanes as well as during the recovery effort. **I am not going to say that social media is going to save your life, but there is a good chance it will help you in some way.**
>
> —Sree Sreenivasan, dean of student affairs,
> Columbia University Graduate School of Journalism

SREE'S TIPS ON USING SOCIAL MEDIA IN
EXTREME WEATHER EMERGENCIES

1. Start accounts with Facebook and Twitter. Most sites have help features to help you get started.
2. Have a list of social media tools and forums you are comfortable with.
3. Know how to take a photograph with a cell phone camera and post it to your social media accounts.
4. Have a way to connect with all your family members via text messaging. (During the quake in Haiti, phone lines were down, so SMS and Twitter were the only ways to communicate.)
5. Group cell phone numbers of family on your phone together so you can access them quickly to make one conference call.
6. Make lists of those among your Facebook friends who are just your family so you can read their specific newsfeeds.
7. Learn geolocation tools, such as FourSquare and Facebook Places, that will help you tell not only what you are doing, but where you are as well.

ON SOCIAL MEDIA TOOLS

66 This is not a toy for kids. This is real stuff that's going to help us. Learn how to use it.

—Sree Sreenivasan (@Sree)

There are lots of ways to upload videos, via your phone or online. Gary Rabkin of VYou (a free social media platform that uses conversational video for web interaction) recommends sharing personal video in emergency situations, if possible, as a means to connect and communicate.[5]

66 Recorded video from a disaster zone can inform people about what is going on, answering questions from other users and sharing critical information. **For example, a user could send video messages about missing people and pets and share information about the status of safety zones, escape routes, and places to purchase supplies and gasoline.**

—Gary Rabkin, VYou

Following relief agencies and government sites on Twitter is also helpful:

66 We interact with people on Twitter in a variety of ways. We use it as a fact-based disaster resource to provide information. We also answer users' questions. We get tweets like "Where can I donate clothes?" and "How do I get help?" We try to answer everyone who contacts us directly.[6]

—Gloria Huang, social engagement specialist for the American Red Cross (@redcross)

SUGGESTED LIST OF EMERGENCY RESOURCES TO FOLLOW ON TWITTER BY @BONNIEWEATHER

@RedCross (look for a link to your local chapter)

_Chapter@FEMA

_Chapter@USNOAAGOV

_Chapter@NHC_ATLANTIC (The National Hurricane Center) and @NHC_PACIFIC

_Chapter@NSSL (The National Severe Storms Laboratory)

_Chapter@NatlParkService

_Chapter@Tsunami_Alarm

_Chapter@Firewise

_Chapter@USGS (earthquake information) and @USGSLive (for live events)

_Chapter@ForestService (USDA Forest service)

_Chapter@FireInfoGirl and @wildfirenews (note: individual states have specialized Twitter handles you can follow)

Also consider following:
- News organizations, both broadcast and cable—including local, regional, and national. You can get the Twitter handles for your favorite local TV stations, newspapers, and magazines as well.
- Meteorologists, national and local.
- Local government—agencies for your town, county, and state.
- Airlines you frequently fly. They may post weather and travel alerts.

CONCLUSION

Social media forums like Facebook and Twitter may be the only sources available with which to communicate in the aftermath of extreme weather and natural disasters. Don't wait until the emergency to figure out how to use these and other helpful social media tools. Before disaster strikes, learn how to post and read status updates, upload pictures and video, and what organizations to follow in order to send and receive information quickly when you need it most.

13

HOW TO CREATE A FAMILY DISASTER PLAN FOR PEOPLE AND PETS

WHAT IS A FAMILY DISASTER PLAN?

A Family Disaster Plan is a personalized action plan that lets each member of a household knows what to do in particular disaster situations and how to be prepared in advance. A functional Family Disaster Plan helps alleviate fears about potential disasters, makes actual disaster situations less stressful, and saves precious time in the face of disasters.

—Source: The Red Cross

WILLIAM CRAIG FUGATE, THE ADMINISTRATOR OF THE Federal Emergency Management Agency (FEMA), encourages families to make a plan for a disaster way ahead of time[1]:

> You won't have time when severe weather hits to talk to your family and figure out what to do. It has to be done in advance.... I call it a family communication plan and this is critical, how are you going to reach each other in the event of an emergency, especially if you are not together when it happens ... and where are you going to go if you have to leave your home.

Ready.gov, a FEMA website, along with the NOAA, has tips on how to make this plan:

THE BASICS OF A FAMILY EMERGENCY PLAN

- Use a NOAA weather radio.[2] Remember to replace its battery every six months, as you do with your smoke detectors.

- Identify an out-of-town contact who could be the point person for separated family members. Phone lines or cell towers may be down in your immediate area. It may be easier to make a long-distance phone call than to call across town.

- Post emergency telephone numbers by your phones and make sure your children know how and when to call 911.

- Be sure every member of your family knows the phone number and has a cell phone and a prepaid phone card to call the emergency contact. If you have a cell phone, program that person(s) as "ICE" (In Case of Emergency) in your phone. If you are in an accident, emergency personnel will often check your ICE listings in order to get hold of someone you know. Make sure to tell your family and friends that you've listed them as emergency contacts.

- Make sure every family member knows how to text message (also known as SMS or Short Message Service). Text messages can often get around network disruptions.

- Subscribe to alert services. Many communities now have systems that will send instant text alerts or e-mails to let you know about bad weather, road closings, and other local emergencies. You can sign up by visiting your local Office of Emergency Management website.

- Periodically check your insurance coverage—flood damage is not usually covered by homeowners insurance.

- Before the disaster, it is a good idea to take first-aid, CPR, and disaster-preparedness classes.

PLANNING TO STAY OR GO

The first important decision is whether to stay where you are or evacuate. You should understand and plan for both possibilities. Use common sense and available information to determine if there is an immediate danger. Local authorities may or may not immediately be able to provide information on what is happening and what you should do. However, you should watch TV, listen to the radio, or check the Internet often for information or official instruction as it becomes available. Remember, the sheriff might not come knocking on your door to tell you when to leave. You might need to determine that yourself.

"Don't wait until the forecast gets better. Leave when you can get out safely with your family. Failure to get out early enough could result in injury or death," says FEMA's Fugate.

Before his position at FEMA, Fugate served as the head of the Florida Division of Emergency Management. In 2004, Fugate managed the largest federal disaster response in Florida history as four damaging hurricanes struck the state in quick succession. Tragically, some Pensacola residents refused to follow mandatory evacuation orders when Hurricane Ivan hit in 2004.

Recalls Fugate:

There are some pretty chilling 911 calls that came in (during Hurricane Ivan). . . . There are people who stayed on the barrier islands and did not evacuate. As the water started rising, and the waves were pounding their homes, they were frantically calling 911 for rescue. And the operators were telling people the wind is too strong and to hang on, but nobody can get to you until the storm passes. . . . People were realizing they had made a horrible mistake. . . . They waited until it was too late, hoping it wasn't going to be that bad, and they were wrong.

The Red Cross details the specific, step-by-step way to implement your Family Disaster Plan.

Once you know what disasters are possible in your area, have a household meeting to talk about how to prepare should one occur. Plan to share responsibilities and to work together as a team.

Know what to do in case household members are separated in a disaster. Disaster situations are stressful and can create confusion. **Keep it simple.**

Pick two places to meet:

1. Right outside your home in case of a sudden emergency, like a fire.
2. Outside your neighborhood in case you cannot return home or are asked to leave your neighborhood.

Choose two out-of-town contacts:

1. A friend or relative who will be your household's **primary** contact.
2. A friend or relative who will be your household's **alternative** contact.

Children should also memorize the primary and alternative contacts' names, addresses, and home and cell telephone numbers, or carry the information with them. Include these contact numbers on your pet's identification tags, or use a national pet-locator service that someone can call if they find your lost pet.

Separation is particularly likely during the day when adults are at work and children are at school.

If household members are separated from one another in a disaster, they should call the primary contact. If the primary contact cannot be reached, they should call the alternative contact. Remember, after a disaster it is often easier to complete a long-distance connection than a local call.

Make sure that adults and children know how to tell the contact where they are, how to reach them, and what happened or to leave this essential information in a brief voice mail.

Discuss what to do if a family member is injured or ill.

Discuss what to do if authorities advise you to evacuate.
Learn about public shelter locations in your community.
Make "in case of evacuation" arrangements for a place
to stay with a friend or relative who lives out of town or

with a hotel, motel, or campground you are familiar with that can be reached by an evacuation route you would expect to take.

Be familiar with evacuation routes. Plan several evacuation routes in case certain roads are blocked or closed. Local officials will direct you to the safest route.

Plan how to take care of your pets. Pets (other than service animals) usually are not permitted in public shelters or other places where food is served. Plan where you would take your pets if you had to go to a public shelter. Many communities are developing emergency animal shelters similar to shelters for people.

Contact your local emergency management agency to find out about emergency animal shelters in your community.

Make and complete a checklist:

- **Post emergency numbers (fire, police, ambulance, etc.) by telephones.** You may not have time in an emergency to look up critical numbers.
- **Teach all responsible members of the household how and when to turn off the water, gas, and electricity at the main switches or valves.** Turn off utilities only if you suspect a leak or damaged lines, or if you are instructed to do so by authorities. If you turn the gas off, you will need a professional to turn it back on. Do not actually turn any valve unless it is a real emergency. Place a tag on shutoff valves to make them easier to identify.
- **Attach a shutoff valve wrench or other special tool in a conspicuous place close to the gas and water shutoff valves.**
- **Check if you have adequate insurance coverage.** Ask your insurance agent to review your current policies to ensure

that they will cover your home and belongings. If you are
a renter, your landlord's insurance does not protect your
personal property; it protects only the building. You need
individual insurance. Flood insurance typically comes from the
National Flood Insurance Program (NFIP), but it can usually
be purchased through the same company that carries your
homeowner's policy. More information on NFIP can be found
online at: http://www.fema.gov/business/nfip/.

- **Learn how to use your fire extinguisher.** Make sure all
  household members know where extinguishers are kept.
  Different extinguishers operate in different ways. There is no
  time to read directions during an emergency. Only adults should
  handle and use extinguishers.

**Practice and maintain your plan to make sure your household is
ready for disaster:**

- Review and update your Family Disaster Plan and your Disaster
  Supplies Kit at least every six months.
- Observe the expiration or "use by" date on stored food and
  water. If you have prepared you own containers of water,
  replace them every six months to ensure freshness.
- Conduct fire and emergency evacuation drills at least twice
  a year. At home, practice escaping from various rooms,
  particularly bedrooms, and meeting at the place you have
  selected right outside your home. Have each driver actually
  drive evacuation routes so each will know the way. Select
  alternative routes and familiarize drivers with them in case the
  main evacuation route is blocked during an actual disaster.
- **Include your pets in your evacuation and sheltering drills.**
  Practice evacuating your pets so they will get used to entering
  and traveling calmly in their carriers. If you have horses or
  other large animals, be sure that they are accustomed to

entering a trailer. Practice bringing your pets indoors, into your safe room, so that if you are required to shelter in place, they will be comfortable.

If you have a disability or a mobility problem, you should consider adding the following steps to the usual preparations:

- **Create a network of relatives, friends, or coworkers to assist in an emergency.** If you think you may need assistance in a disaster, discuss your disability with relatives, friends, or coworkers and ask for their help. For example, if you need help moving or getting necessary prescriptions, food, or other essentials, or if you require special arrangements to receive emergency messages, make a plan with friends or helpers.

- **Make sure they know where you keep your Disaster Supplies Kit.** Give a key to a neighbor or friend who may be able to assist you in a disaster.

- **Maintain a list of important items and store the list with your Disaster Supplies Kit.** Give a copy to another member of your household and a friend or neighbor. Include things that might be counterintuitive about your health:

  - Special equipment and supplies—for example, hearing aid batteries.

  - Current prescription names, sources, and dosages.

  - Names, addresses, and telephone numbers of doctors and pharmacists. If you get prescriptions by mail, confirm where you will be able to get them locally in an emergency.

  - Detailed information about the specifications of your medication or medical regimen, including a list of things incompatible with medication you use—for example, aspirin.

- **Contact your local emergency management office.** Many local emergency management offices maintain registers of people with disabilities and their needs so they can be located and assisted quickly in a disaster.

- **Put on medical alert tags or bracelets to identify your disability** at the first sign of weather disaster. These may save your life if you are in need of medical attention and unable to communicate.
- **Know the location and availability of more than one facility if you are dependent on a dialysis machine or other life-sustaining equipment or treatment.** There may be other people requiring equipment, or facilities may have been affected by the disaster.

If you have a severe speech, language, or hearing disability:

- When you dial 911 (or your local emergency number), tap the space bar to indicate a TDD (Telecommunications Device for the Deaf) call.
- Store a writing pad and pencils to communicate with others.
- Keep a flashlight handy to signal your whereabouts to other people and for illumination to aid in communication.
- Remind friends that you cannot completely hear warnings or emergency instructions. Ask them to be your source of emergency information as it comes over the radio. Another option is to use a NOAA weather radio with an alert feature connected to a light. If a watch or warning is issued for your area, the light will alert you to potential danger.
- If you have a hearing ear dog, be aware that the dog may become confused or disoriented in an emergency. Store extra food, water, and supplies for your dog. Trained hearing ear dogs will be allowed to stay in emergency shelters with their owners. Check with local emergency management or American Red Cross, Salvation Army, or other shelter officials for more information.

If you have a service animal:

- Be aware that the animal may become confused or disoriented in an emergency. Disasters may often mask or confuse scent markers that are part of your service animal's normal means of navigation.
- If you are blind or visually impaired, keep extra canes around your home and office, even if you use a guide dog.
- If you have a guide dog, train the dog to know one or two alternate routes. A guide dog familiar with the building may help you and others find a way out when no one else can see.
- Be sure your service animal has identification and your phone numbers attached to its collar, including emergency contact information through a national pet locator service.
- Trained service animals will be allowed to stay in emergency shelters with their owners. Check with your local emergency management agency or American Red Cross, Salvation Army, or other shelter officials for more information.

If you use a wheelchair:

- Show friends how to operate your wheelchair or help you transfer out of your chair so they can move you quickly if necessary.
- If you use a power wheelchair, make sure friends know the size of your wheelchair, in case it has to be transported, and know where to get a battery if needed.
- Make arrangements with others to be carried out, if necessary, and practice doing that.

FOR THE ELDERLY (TIPS FROM THE CALIFORNIA EMERGENCY MANAGEMENT AGENCY)

Find two people you trust who will check on you after an emergency. Tell them your special needs. Show them how to operate

any equipment you use. **Show them where your emergency supplies are kept. Give them a spare key.**

- Make it as easy as possible to quickly get under a sturdy table or desk for protection during an earthquake or explosion.
- Anchor special equipment, such as telephones and life-support systems. Fasten tanks of gas, such as oxygen, to the wall.
- Keep a list of medications, allergies, special equipment, names, and numbers of doctors, pharmacists, and family members. Make sure you have this list with you or that it's easily accessible.
- Keep an extra pair of eyeglasses and medication with your emergency supplies.
- Keep walking aids near you at all times. Have extra walking aids in different rooms of the house.
- Put a security light in each room. These lights plug into any outlet and light up automatically if there is a loss of electricity. They continue operating automatically for four to six hours, and they can be turned off by hand in an emergency.
- Make sure you have a whistle to signal for help.
- Keep extra batteries for hearing aids with your emergency supplies. Remember to replace them annually.
- Prepare to be self-sufficient for at least three days.
- Turn on your portable radio for instructions and news reports. For your own safety, cooperate fully with public safety officials and instructions.
- If you evacuate, leave a message at your home giving family members your location.

PLAN FOR YOUR PETS
(FEMA, Humane Society, American Red Cross, and ASPCA)

If you have pets, you should:

- **Take your pets with you if you evacuate.** If it is not safe for you, it is not safe for them. Leaving them may endanger you, your pets, and emergency responders.

- **Plan in advance where you will go if you evacuate,** as pets (other than service animals) are usually not allowed in public shelters.

- **Contact hotels and motels outside your immediate area to check their policies on accepting pets** and restrictions on the number, size, and species. Ask if "no pet" policies could be waived in an emergency.

- **Ask friends, relatives, or others outside your area if they could shelter your animals.** If you have two or more pets, they may be more comfortable if kept together, but be prepared to house them separately.

- **Prepare a list of boarding facilities and veterinarians who could shelter animals in an emergency;** include 24-hour telephone numbers. Ask local animal shelters if they provide emergency shelter or foster care for pets in a disaster situation. Animal shelters may be overburdened, so this should be your last resort unless you make such arrangements well in advance.

- **Keep a list of "pet friendly" places, including their telephone numbers, with other disaster information and supplies.** If you have notice of an impending disaster, call ahead for reservations. Hotels and motels with no-pet policies may waive these policies during a disaster, particularly if the pet is housed in a carrier. Contact establishments along your evacuation route to see if they will waive no-pet rules, and make sure you have adequate facilities and supplies for your pets.

- **Carry pets in a sturdy carrier.** Animals may feel threatened by some disasters, become frightened, and try to run. Being in its own carrier helps reassure a pet.

- **Have identification, collar, leash, and proof of vaccinations for all pets.** At some locations, you may need to provide veterinary records before boarding your pets. If your pet is lost, identification will help officials return it to you.

- **Assemble a portable pet disaster supplies kit.** Keep food, water, and any special pet needs in an easy-to-carry container. More information on this can be found in chapter 15.
- **Have a current photo of your pets** in case they get lost.

Create a plan in case you are not at home during an emergency to ensure that someone takes care of your pets, even evacuating them if necessary. The plan should include these elements:

- Give a trusted neighbor the key to your home and instructions, as well as your daytime (work or school) contact information.
- Make sure the neighbor is familiar with your pets and knows the location of your pet emergency kit.
- Make sure the neighbor listens to a local radio or television station for emergency information and puts your shelter-in-place or pet evacuation plan into action.
- Have a plan to communicate with your neighbor after the event. You will want to arrange a meeting place in a safe area so you can be reunited with your pets.
- Contact your local emergency management agency, humane society, and animal control agency to see if your community has sheltering options for animals and for families with pets. If not, learn more about emergency animal shelters and volunteer to include this option in local disaster-preparedness efforts.
- Learn pet first aid and keep your pet first-aid kit up-to-date.
- (A complete list of Emergency Supplies for Pets can be found in Chapter 15.)

14

EMERGENCY SUPPLIES FOR PEOPLE

IT'S POSSIBLE THAT IN THE COURSE OF ANY WEATHER emergency or natural disaster, you may lose power for days, weeks, or even longer. That's why it's important to have your emergency disaster supplies stocked, ready, and easily accessible. Before you shop for everything, check with family members to see what supplies you already have. Also, periodically go over your supplies to make sure they are useable and up to date. Because you might not have access to your phone, computer, or a portable reader in the time of a natural disaster, **always have your NOAA weather radio ready (with working batteries) so you can receive weather updates or evacuation orders.** Make sure all family members know where the radio is and how to work it.

WHAT TO HAVE IN BASIC DISASTER SUPPLIES AT HOME

The following items are recommended by FEMA's Ready.gov and the American Red Cross for inclusion in your disaster supplies checklist.

✓ At least a three-day supply of water—one gallon of water per person, per day for at least three days. Note that the American Red Cross adds: "Store *at least* a 3-day supply and consider storing a two-week supply of water for each member of your family. If you are unable to store this much, store as much as you can. . . . You will also need water for food preparation and hygiene."

✓ At least a three-day supply of non-perishable food per person, per day

✓ Portable, battery-powered or hand-cranked radio or television and a NOAA Weather Radio with tone alert and extra batteries

✓ Flashlight and extra batteries

✓ First-aid kit and manual. Make sure it is fully stocked and nothing inside it is expired.

✓ Copies of personal documents (medication list and pertinent medical information, proof of address, deed/lease to home, passports, birth certificates, credit card and bank information, insurance policies) secured in a waterproof container

✓ Family and emergency contact information accessible to everyone in the home, also stored in a waterproof container

✓ Sanitation and hygiene items (moist towelettes and toilet paper)

✓ Matches and waterproof container

✓ Dust mask, to help filter contaminated air and plastic sheeting and duct tape to shelter in place

✓ Whistle

✓ Extra clothing

✓ Cell phone with extra chargers

✓ Can opener for food

✓ Wrench or pliers

✓ Cash and coins. You may need this in case credit cards aren't accepted or an ATM is not accessible.

✓ Matches in a waterproof container

✓ Local maps

✓ Moist towelettes, garbage bags, and plastic ties

✓ Fire extinguisher

✓ Special-needs items, such as prescription medications, eyeglasses, contact lens solutions, and hearing aid batteries

✓ Items for infants, such as formula, diapers, bottles, and pacifiers

✓ Towels

✓ Work gloves

- ✓ Tools/supplies for securing your home
- ✓ Plastic sheeting
- ✓ Duct tape
- ✓ Scissors
- ✓ Household liquid bleach (unscented, *not* color safe, or bleaches with added cleaners)
- ✓ An eyedropper
- ✓ Paper plates, cups, plastic utensils, paper towels
- ✓ Feminine hygiene supplies
- ✓ Sand, rock salt, or non-clumping kitty litter to make walkways and steps less slippery
- ✓ Entertainment items: books, games, puzzles, or activities for children

It's advised that you take into consideration where you live and the type of extreme weather your home is susceptible to. For example, FEMA advises that if you live in a cold climate, you must think about ways to stay warm. In an ice storm or blizzard, it is possible that you will not have heat. Think about your clothing and bedding supplies. Be sure to include one complete change of clothing and shoes per person, including[1]:

- ✓ Jacket or coat
- ✓ Long pants
- ✓ Long-sleeve shirt
- ✓ Sturdy shoes
- ✓ Hat, mittens, and scarf
- ✓ Sleeping bag or warm blanket (per person)

Remember:

1. Be sure to account for growing children and other family changes. Also take into account any family member with special needs.

2. When you put your supplies together, make sure to maintain them so they are all safe to use when needed.

FEMA'S TIPS ON WATER STORAGE FOR EMERGENCIES

To prepare the safest and most reliable emergency supply of water, it is recommended that you purchase commercially bottled water. Keep bottled water in its original container and do not open it until you need to use it. If you are preparing your own containers of water, it is recommended that you purchase food-grade water storage containers from surplus or camping supplies stores. Before filling, thoroughly clean the containers with dishwashing soap and water, and rinse completely. If you choose to use your own storage containers, choose two-liter plastic soft-drink bottles—not plastic jugs or cardboard containers that have had milk or fruit juice in them. Milk protein and fruit sugars cannot be adequately removed from these containers and provide an environment for bacterial growth when water is stored in them. Cardboard containers also leak easily and are not designed for long-term storage of liquids. Also, do not use glass containers, because they can break and are heavy.

Filling water containers: Fill the bottle to the top with tap water. If the tap water has been commercially treated from a water utility with chlorine, you do not need to add anything else to the water to keep it clean. Tightly close the container using the original cap. Be careful not to contaminate the cap by touching the inside of it with your finger. Place a date on the outside of the container so that you know when you filled it. Store in a cool, dark place. Replace the water every six months if you are not using commercially bottled water. If the water you are using comes from a well or water source that is not treated with chlorine, Ready.gov advises: "In an emergency, you can use household chlorine bleach to treat by using 16 drops of regular household liquid bleach per gallon of water. Do not use scented, color safe or bleaches with added cleaners."

FEMA'S TIPS ON WHAT KINDS OF FOOD TO STORE

Avoid foods that will make you thirsty. Choose salt-free crackers, whole grain cereals, canned foods with high liquid content, stock dry mixes, and other staples that do not require refrigeration, cooking, water, or special preparation. Include special dietary needs. **(Note: Be sure to include a manual can opener.)**

THE AMERICAN RED CROSS'S GUIDE ON EMERGENCY FOOD STORAGE

Storage Tips:

- Keep food in a dry, cool spot—a dark area if possible.
- Open food boxes and other re-sealable containers carefully so that you can close them tightly after each use.
- Wrap perishable foods, such as cookies and crackers, in plastic bags and keep them in sealed containers.
- Empty open packages of sugar, dried fruits, and nuts into screw-top jars or air-tight canisters for protection from pests.
- Inspect all food for signs of spoilage before use.
- Throw out canned goods that become swollen, dented, or corroded.
- Use foods before they go bad, and replace them with fresh supplies, dated with ink or marker. Place new items at the back of the storage area and older ones in front.

WHAT TO HAVE IN YOUR OFFICE

Many offices have first-aid kits in them already. Here are the items the kits should contain, according to the Red Cross[2]:

✓ If your employer does not provide these first-aid supplies, have the following essentials on hand, if possible:

✓ (20) adhesive bandages, various sizes

- ✓ 5" × 9" sterile dressing
- ✓ conforming roller gauze bandage
- ✓ triangular bandages
- ✓ 3" × 3" sterile gauze pads
- ✓ 4" × 4" sterile gauze pads
- ✓ roll 3" cohesive bandage
- ✓ germicidal hand wipes or waterless alcohol-based hand sanitizer
- ✓ 1 pack of (6) antiseptic wipes
- ✓ pair large medical grade non-latex gloves
- ✓ adhesive tape, 2" width
- ✓ antibacterial ointment
- ✓ cold pack
- ✓ scissors (small, personal)
- ✓ tweezers
- ✓ CPR breathing barrier, such as a face shield

Other emergency safety items to have at the office:

- ✓ flashlight with extra batteries
- ✓ battery-powered radio
- ✓ Food—enough nonperishable food to sustain you for at least one day (three meals) is suggested. Select foods that require no refrigeration, preparation, or cooking, and little or no water. The following items are suggested: ready-to-eat canned meals, meats, fruits, and vegetables, canned juices, and high-energy foods (granola bars, energy bars, etc.). Don't forget a nonelectric can opener.
- ✓ Water—keep at least one gallon available, or more if you are on medications that require water or that increase thirst. Store water in plastic containers such as soft-drink bottles. Avoid using containers that will decompose or break, such as milk cartons or glass bottles. (See FEMA's tips above on how to store water.)

✓ Medications—include usual nonprescription medications that you take, including pain relievers, stomach remedies, etc. If you use prescription medications, keep at least three days' supply of these medications at your workplace. Consult with your physician or pharmacist about how these medications should be stored, and with your employer about storage concerns.

In the unlikely event that extreme weather or natural disaster causes you to have to spend the night at your workplace, you may want to consider also having these supplies on hand at the office:

✓ emergency blanket (Mylar is recommended for warmth)
✓ paper plates and cups, plastic utensils
✓ personal hygiene items, including a toothbrush, toothpaste, comb, brush, soap, contact lens supplies, and feminine supplies
✓ plastic garbage bags, ties (for personal sanitation uses)
✓ at least one complete change of clothing and footwear, including a long-sleeved shirt and long pants, as well as closed-toed shoes or boots
✓ if you wear glasses, keep an extra pair with your workplace disaster supplies.

WHAT EMERGENCY SUPPLIES TO KEEP AS A KIT IN YOUR CAR

Edmonds.com describes the emergency items below as the "basic ones" to keep in your car. Some of the basic items include[3]:

✓ 12-foot jumper cables
✓ four 15-minute roadside flares
✓ two quarts of oil
✓ gallon of antifreeze

✓ first-aid kit (including an assortment of bandages, gauze, adhesive tape, antiseptic cream, instant ice and heat compresses, scissors and aspirin)

✓ blanket

✓ extra fuses

✓ flashlight and extra batteries

✓ flat head screwdrivers

✓ Phillips head screwdrivers

✓ pliers

✓ vise grips

✓ adjustable wrench

✓ tire inflator (such as a Fix-A-Flat)

✓ tire pressure gauge

✓ rags

✓ roll of paper towels

✓ roll of duct tape

✓ spray bottle with washer fluid

✓ pocketknife

✓ ice scraper

✓ pen and paper

✓ help sign

✓ granola or energy bars

✓ bottled water

✓ heavy-duty nylon bag to carry it all in

NOTE: Ricky Slack, a member of Tennessee's Swift Water Rescue team recommends that you consider having a "window punch" or "life hammer" tool accessible in your car. This device would allow you to break a window easily from the inside if you are ever trapped in your car and need to escape.

For emergency supply kits, consider the weather environment where you live. If you experience cold winters, for example, you will want to have additional items handy to keep warm in an emergency. No one knows this better than Staff Sergeant Mark Dornford, who teaches arctic survival techniques to defense personnel in the below-zero conditions of Moose Creek, Alaska:

> If you live or are working in an area with cold temps and lots of snow, you should have a good sleeping bag, extra warm clothes, food, water [or a way to melt snow—learn how to do that in Chapter 10], a metal container, candles, matches, winter hat, gloves, boots, flashlight, and some sort of signal light. I have all this stuff packed into a duffel bag that I keep in the back of my car.[4]

15
EMERGENCY SUPPLIES FOR PETS

WHAT TO HAVE AT HOME FOR PETS

The Red Cross advises that you "keep items in an accessible place and store them in sturdy containers."[1]

- ✓ phone number and addresses: veterinarian, emergency vet, poison control
- ✓ a basic pet first-aid book
- ✓ photocopies of important documents: medical records, vaccinations, proof of ownership, contact info for veterinarian, a picture of your pet
- ✓ a transport carrier that the pet can stand and turn around in (it may have to live in it temporarily during evacuation) and a leash
- ✓ supply of food, water, litter, litter box for three to seven days
- ✓ medical gloves: to protect hands and prevent contamination
- ✓ scissors: to cut gauze or the animal's hair
- ✓ bottled water
- ✓ a mild antibacterial soap: to clean skin and wounds
- ✓ paper towels
- ✓ gauze pads: for wounds
- ✓ gauze rolls: for wounds, and can also be used as a temporary muzzle
- ✓ alcohol prep pads: to sterilize equipment—NOT for use on wounds

✓ self-adhesive bandages: flexible bandage used to wrap and stabilize injuries (do not wrap too tightly)

✓ a large cloth towel: to wrap animal in

✓ iodine ointment: for burns, cuts, and abrasions

✓ eyewash: like contact lens solution or water in a squeeze bottle to gently but thoroughly flush out wounds and eyes

✓ triple antibiotic ointment: for cuts and abrasions (never for eyes)

✓ cotton swabs

✓ tweezers: for the removal of foreign objects from skin and paws

✓ special tick tweezers: for the proper removal of ticks

✓ pet bed or toys if easily transportable

IF YOU HAVE TO EVACUATE WITH YOUR PET

The ASPCA advises that you "arrange a safe haven for your pets in the event of evacuation.[2] DO NOT LEAVE YOUR PETS BEHIND. Remember, if it isn't safe for you, it isn't safe for your pets. They may become trapped or escape and be exposed to numerous life-threatening hazards. **Note that not all Red Cross, Salvation Army, or other disaster shelters accept pets**, so it is imperative that you have determined where you will bring your pets ahead of time."

Before you have to evacuate, the ASPCA recommends microchipping your pet as a more permanent form of identification. A microchip is implanted in the animal's shoulder area, and can be read by scanner at most animal shelters.

The ASPCA also suggests that you take care of these additional tasks before a disaster strikes so you are prepared:

✓ Contact your veterinarian for a list of preferred boarding kennels and facilities.

✓ Ask your local animal shelter if they provide emergency shelter or foster care for pets.

✓ Identify hotels or motels outside your immediate area that accept pets.

✓ Ask friends and relatives outside your immediate area if they would be willing to take in your pet.

WHAT TO TAKE WITH YOU IF YOU HAVE TO EVACUATE WITH YOUR PET

The Humane Society says you should keep an Evac-Pack and supplies handy for your pets. Make sure that everyone in the family knows where it is. This kit should be clearly labeled and easy to carry. Items to consider keeping in or near your pack include:

✓ pet first-aid kit

✓ three to seven days' worth of canned (pop-top) or dry food (be sure to rotate every two months)

✓ disposable litter trays (aluminum roasting pans are perfect)

✓ litter or paper toweling

✓ liquid dish soap and disinfectant

✓ disposable garbage bags for cleanup

✓ pet feeding dishes

✓ extra harness and leash (note: harnesses are recommended for safety and security)

✓ photocopies of medical records and a waterproof container with a two-week supply of any medicine your pet requires (write the date and check for expiration)

✓ bottled water, at least seven days' worth for each person and pet (store in a cool, dry place and replace every two months)

✓ a traveling bag, crate, or sturdy carrier, ideally one for each pet

✓ flashlight

✓ blanket (for scooping up a nervous pet)

✓ recent photos of your pets (in case you are separated and need to make "lost" posters)

✓ especially for cats: pillowcase or toys, scoopable litter

✓ especially for dogs: long leash and yard stake, toys and chew toys, a week's worth of cage liner

WHAT TO KEEP ON HAND TO EVACUATE OTHER TYPES OF ANIMALS

The ASPCA says that special consideration needs to be taken for other types of pets[3]:

For Birds

✓ Birds should be transported in a secure travel cage or carrier.

✓ In cold weather, make certain you have a blanket over your bird's cage. This may also help reduce the stress of traveling.

✓ In warm weather, carry a spray bottle to periodically moisten your bird's feathers.

✓ Have recent photos available, and keep your bird's leg bands on for identification.

✓ If the carrier does not have a perch, line it with paper towels that you can change frequently.

✓ Keep the carrier in as quiet an area as possible.

✓ It is particularly imperative that birds eat on a daily basis, so purchase a timed feeder. If you need to leave your bird unexpectedly, the feeder will ensure his daily feeding schedule.

✓ Items to keep on hand: catch net, heavy towel, blanket or sheet to cover cage, cage liner.

For Reptiles

✓ A snake may be transported in a pillowcase, but you should have permanent and secure housing for him when you reach a safe place.

✓ Take a sturdy bowl that is large for your reptile to soak in. It's also a good idea to bring along a heating pad or other warming device, such as a hot water bottle.

Lizards can be transported like birds (see p. 226).

For Other Small Animals
- ✓ Small animals, such as hamsters, gerbils, mice, and guinea pigs, should be transported in secure carriers with bedding materials, food, and food bowls.
- ✓ Items to keep on hand: salt lick, extra water bottle, small hide box or tube, and a week's worth of bedding.

WHAT EMERGENCY SUPPLIES TO KEEP IN YOUR CAR FOR PETS

It's always a good idea to keep extra food and water and a small first-aid kit containing basic medical supplies for your pet in your car.

When you travel by car with your pet, be aware of the temperature. Animals are particularly sensitive to heat. According to veterinarians Dr. Charlotte Krugler and Dr. Adam Eichelberger of Clemson University, "**The most serious mistake is to leave a pet in a closed automobile on a hot day. Leaving the windows open will not prevent the temperature of the car, and then the pet, to escalate. Many of these pets, even with rigorous treatment, do not survive, which is very sad since so preventable.**"[4]

CONCLUSION

Pets are part of our family but also have their own singular needs, and it's important to prepare separately for them. Have your pet emergency supplies ready, up-to-date, and accessible at home. Make plans for how you would evacuate your pet and what supplies you would take with you in advance. Keep in mind your pet's special needs, like keeping them comfortable in extreme hot or cold weather, during the journey to a safer environment.

ACKNOWLEDGMENTS

THERE ARE MANY PEOPLE AND ORGANIZATIONS I WOULD like to thank for making this book possible. First and foremost, I feel grateful and privileged to all the people I interviewed personally, who shared their incredible survivor stories with me and with my readers: Mary Theriot, Sandee LaMotte, Shaun Gholston, Nicole and Luis, Bill Harbison, Rick Van Feldt, Essie Hendrix, Rick Jarvis, Steve Chamberlin, and Todd Archer. The courage you, and the other disaster survivors profiled in this book, displayed is truly inspirational.

I would like to thank the many government and private agencies, organizations, and universities for their resources in science, safety, and emergency advice: NOAA and FEMA, the USGS, the Centers for Disease Control, the American Red Cross, the ASPCA, the Humane Society, the United States Air force, Eielson Air Force Base, Pacific Air Forces, the Skip Barber Racing School, Georgia Tech, the U.S. Lifesaving Association, the Los Angeles County Fire Department, the Rocky Mountain Fire Department, the Florida Environmental Protection Agency, the Lightning Safety Institute, the Dallas Cowboys, the North Shore Animal League, Clemson University Livestock-Poultry Health, the National Hurricane Center, the National Severe Storms Laboratory (special mention to the now-retired Don Burgess), the office of William Craig Fugate (FEMA), the U.S. Park Rangers, Zion and Yellowstone National Parks, the California Emergency Management Agency, Kansas State University Pet Health Center, the EPA, Sonoma County Animal Control, the Smith County (Tennessee) Emergency Services SWIFT Water Rescue Team, the National Center for Atmospheric Research, the Pacific Tsunami Warning Center, the Japanese Meteorological Agency, the National Center for Atmospheric Research, the University of Washington Department of Earth and Space Sciences, the Department of Earth and Environmental Sciences of Tulane University, Colleen Paige,

PetSitters.com, Facebook, VYou, Twitter, Columbia University, and the International Fund for Animal Welfare.

I would like to pay a specific tribute to an expert quoted in the book, the late Dr. Kurt Frankel. Kurt was an assistant professor at Georgia Tech's School of Atmospheric Science and was often featured on-air as an earthquake expert on CNN. He was readily available to answer all my interview questions for this book in the early part of 2011. Kurt was tragically killed when a motorist struck the bicycle he was riding in July 2011. His legacy of knowledge continues through his students. The School of Earth and Atmospheric Sciences at Georgia Tech is establishing an endowment fund in Kurt Frankel's name to recognize and support outstanding students in solid earth geosciences. For more information, contact:

Laura Cederquist
School of Earth and Atmospheric Sciences
Georgia Institute of Technology
311 Ferst Drive
Atlanta GA 30332–0340

I am so pleased that Luba Ostashevsky at Palgrave Macmillan was my editor for this book. Luba is intelligent and professional. I feel very fortunate that I had her expertise and input for this book. I would like to thank all the people I worked with at Macmillan: Laura Lancaster in Editorial, Siobhan Paganelli in Publicity and Michelle Fitzgerald in Marketing, Donna Cherry in Production, and the entire sales team.

My television career has been a long and rewarding journey. I would like to thank the people who gave me my start in the early years, Jim Serra of KPLC-TV and Pat Dolan of News 12 Long Island, as well as the many who make my experience at CNN and HLN so fulfilling today. At CNN, I'm privileged to work with an outstanding group of meteorologists. I'd like to make special mention of two: Rob Marciano, with whom, as he would say, "we go way back." Rob and I both started our careers at KPLC-TV in Lake Charles, Louisiana, years ago. It was nice to see a familiar face when I started at CNN in 2005.

I also want to mention my colleague Chad Myers. Through my years here, Chad has been not only been a great friend and coworker, but also a mentor to me as well. My supervisor, Marylynn Ryan, was so happy for me when I first told her about this book and offered immediate support. CNN executives Jim Walton, Ken Jautz, Jack Womack, Tony Maddox,

Parisa Khosravi, Rick Davis, as well as HLN's Scot Safon, have all been extremely supportive of this project.

Also at CNN, I would like to thank all the talented producers and all the staff I am privileged to work with. In marketing, Kate Swenson, and in public relations, Bridget Leninger in Atlanta and Carolyn Disbrow in New York, were all terrific in helping me, along with the fantastic sales and marketing team at Macmillan, promote this book.

Former CNNers who were instrumental in helping me grow, as well as offering me career guidance, include Sue Bunda, Peter Dykstra, Sandy Malcolm, and Henry Mauldin.

Another former staffer, Angela Fritz, also offered me her expertise and her encouragement for this project. I'd also like to thank CNN's Debora Chambers for her helpful assistance.

I'm lucky to have a fabulous literary agent named J. L. Stermer. Her enthusiasm for this book from the beginning was terrific and motivating. J. L., along with Paul Fedorko, helped me get the creative process going early on. I found all our conversations about this book and what it could be very inspiring.

My television agent, Adam Leibner, has represented me for a long time. Adam is not only a fantastic and talented agent, he's also a great person. He's always been there for the ups and downs, and has never stopped believing in me. I also want to thank everyone else at NS Bienstock, with a special mention to Carole Cooper, for all their support throughout my career.

My dream to write a book was one I've had since childhood. I often sought the advice of successful authors for guidance. I interviewed *New York Times* best-selling author Nelson DeMille for my former TV station, News 12 Long Island, back in 2000. He was kind enough to answer my many questions and allow me to stay in touch through the years. It's been my pleasure to read all his books and to get to see him in person at his sold-out book signings when he comes to Atlanta.

A few years ago, at a dinner in Atlanta, my friend Rhonda Wise was thoughtful enough to seat me next to *New York Times* best-selling author Jeff Zaslow. He listened as I told him about the idea for my book and gave me valuable insight into the publishing process.

My friend Lisa Bloom, who is also a *New York Times* best-selling author, was also very encouraging to me. I admire her passion and success in all she does.

During Hurricane Katrina in 2005, I had the privilege of interviewing former director of the National Hurricane Center Max Mayfield on

CNN. I know Max's time is extremely valuable, and I am so appreciative and honored that he agreed to write the foreword for this book.

In the process of writing this book, I spent numerous hours alone at my computer, and thus, I disconnected from many of my close friends. I thank them (Shira Sky, Susan Hendricks, Fredricka Whitfield, Bryce Gruber, Beth Weitzman, Mandy Carranza, Rhonda Wise, Craig Setzer, Samantha Parsons, Jackie Naggar, Felice Desner, Dawn Knauer Tresh, Vanessa Vinci, Kimberly Kaloz, and my old friend Andrew Ehinger) for their support, encouragement, and patience with me.

Finally, I want to thank my loving family: My mother's husband Roger, whose Stuart Computer Services saved my computer from crashing many times. He also kindly printed out pages and pages of this manuscript for my mom to read. Roger's encouraging words and support along the way were always so appreciated. My father's wife Andrea not only offered me encouraging words in our phone calls, but she also provided me feedback with my book proposal in the early stages. When she got excited about a chapter I was working on, I got more excited about it! Andrea also is an avid reader, so her opinions on the pacing of each story I told were extremely helpful. I want to thank Andrea and Roger for their support every step of the way.

I was fortunate to grow up with a family that encouraged me to believe in myself from a very early age. My grandparents were actively in my life into my adulthood. I have nothing but fond memories of their smiling faces. It is how I think of them now. They would have really loved to read this book.

When I look back on my life, the past and the present, there are three people who were always there with me. My sister Karyn is not only my little sister, but also my best friend. I don't know how many times we speak a day, but suffice it to say, it's often.

I grew up with my mother Judy's praise and encouragement in my ear. That gave me the self-confidence I needed to pursue all of my endeavors, including a tough career. My mother taught me to stay positive and to be a good person. She also is a retired teacher who is excellent at grammar and spelling. Thank you, Mom, for helping me edit, too!

When I was little, I used to follow my dad, Jerry, around the house. I wanted to do whatever he was doing, go wherever he was going, because I loved and admired him so much. Well, many years later, I still feel that way. There is no one I look up to more than my father. He taught me to work hard and always do my best. As a grown woman today, I am so grateful and appreciative of my dad's unwavering support of everything I'm doing and that he still tells me he's there for me 24/7.

I am very fortunate to have such wonderful parents.

Lastly, I would like to offer a special word of encouragement to any young woman reading this who is pursuing her goals and going through a difficult or challenging time: Remember that with every setback comes an opportunity to revise and rebuild. Rejection is also part of the process. Stay focused, seek mentors, embrace constructive criticism, and you will get there.

When you believe in yourself, anything is possible.

Bonnie Schnoel

NOTES

INTRODUCTION

1. United Nations International Strategy for Disaster Reduction, "Killer Year Caps Deadly Decade—Reducing Disaster Impact Is 'Critical' Says Top UN Disaster Official," January 24, 2011, http://www.unisdr.org/archive/17613.

CHAPTER 1: HURRICANES

1. Author Interview with Mary Theriot, August 2009.
2. Saffir-Simpson Hurricane Wind Scale, http://www.nhc.noaa.gov/sshws .shtml.
3. "Hurricane Basics," http://www.nhc.noaa.gov/HAW2/english/basics.shtml.
4. Author Interview with Dr. Charlotte Krugler, February 2011.

CHAPTER 2: TORNADOES

1. Author Interview with Essie Hendrix, March 2011.
2. "The April 23–24, 2010 Tornado Outbreak," http://www.srh.noaa.gov /jan/?n=2010_04_24_main_tor_madison_parish_oktibbeha.
3. Kenneth Podrazik and Aubry Wilkins, eds., *The Weather Whisper* (Des Moines, IA: National Weather Service, 2011).
4. Ibid.
5. "Tornado Basics," http://www.nssl.noaa.gov/primer/tornado/tor_basics .html.
6. Ibid.
7. "The Online Tornado FAQ," http://www.spc.noaa.gov/faq/tornado/.
8. Author interview with Dr. Scott Mason, February 2011.

CHAPTER 3: FLASH FLOODS

1. Andrew DeMillo and Chuck Bartels, "Death Toll at 18 in Arkansas Flash Flooding; 22 Still Missing," *Commercial Appeal,* June 12, 2010.
2. Ibid.
3. Susan Lindsey, "Nine of Dead Had Ties to Texarkana Area," *Texarkana Gazette Sentinel-Record,* June 15, 2010, http://www.hotsr .com/news/WireHeadlines/2010/06/15/nine-of-dead-had-ties-to-texarkana -area-67.php; Chuck Bartels, "Police Say 20th Body Found in Flooding Disaster," *Times Record* (Fort Smith, AK), June 14, 2010, http://www.swtimes.com/week-in-review/news/article_d23b472c-77

d0-11df-bfbe-001cc4c002e0.html, or http://www.msnbc.msn.com/id/3768 1729/ns/us_news-life/#.Tj1SYGPczUY.

4. Bartels, "Police Say 20th Body Found in Flooding Disaster."
5. "Death Toll Hits 19 in Arkansas Floods," MSNBC.com, June 13, 2010, http://www.msnbc.msn.com/id/37637416/ns/us_news-life/t/death-toll-hits -arkansas-floods/#.Tj1RSWPczUY.
6. Lindsey, "Nine of Dead Had Ties to Texarkana Area."
7. Author Interview with Craig Thexton.
8. Podcast, Kansas State University, http://www.k-state.edu/media/audio/pod casts/Disasterprep.mov.

CHAPTER 4: EXTREME HEAT

1. Name changed for privacy.
2. Author interview with Rick Jarvis.
3. Author interview with Drs. Charlotte Krugler and Adam Eichelberger.

CHAPTER 5: WILDFIRES

1. Tanya Tucker audio clip interview with reporter Brooke Anderson, *Showbiz Tonight,* October 23, 2007, transcript at cnn.com, http://transcripts.cnn .com/TRANSCRIPTS/0710/22/sbt.01.html.
2. "Tanya Tucker and Media Works Seal Deal on New Reality Series," http: //www.mwctv.com/production/2007/08/tanya-tucker-mwc-reality-show/.
3. "Tanya Tucker's Family Evacuated from Malibu Home," GACTV.com, http: //www.gactv.com/gac/nw_headlines/article/0,,GAC_26063_5728438,00 .html.
4. Author interview with Rich Thompson, National Weather Service Forecast Office, Los Angeles/Oxnard, March 18, 2011.
5. "Total Wildland Fires and Acres (1960-2009)," National Interagency Fire Center, http://www.nifc.gov/fireInfo/fireInfo_stats_totalFires.html.
6. Author interview with Rich Thompson, National Weather Service Forecast Office, Los Angeles/Oxnard. March 18, 2011.
7. Author interview with Don Whittemore, Asst. Fire Chief, Rocky Mountain Fire, February 23, 2004.
8. "Wildfire Safety Tips," http://www.ready.gov/be-informed.
9. Ibid.
10. Colleenpaige.com, http://www.pethomemagazine.com/e_petparedness.htm.
11. Ibid.
12. Ibid.
13. "Pet Fire Safety," Veterinary Pet Insurance, A Nationwide Company, http:// www.petinsurance.com/healthzone/pet-articles/pet-owner-topics/Pet-Fire -Safety.aspx.
14. Ibid.
15. Ibid.

CHAPTER 6: RIP CURRENTS

1. Author interview with Sandee LaMotte, March 27, 2011.
2. Julie Hauserman, "Currents Deadly for Florida Tourists," *St. Petersberg Times,* July 20, 2003, http://www.sptimes.com/2003/07/20/State/Currents _deadly_for_F.shtml.

3. Ibid.
4. "Rip Current Safety," NOAA website, http://www.ripcurrents.noaa.gov/.
5. Ibid.
6. Author Interview with B. Chris Brewster, March 30, 2011.
7. "Rip Current Safety."
8. "Beach Warning Flag Program," Florida Dept. of Environmental Protection website, http://www.dep.state.fl.us/cmp/programs/flags.htm.
9. Ibid.
10. Author Interview with B. Chris Brewster.
11. Daniel J. DeNoon, "Rip Current No. 1 Beach Danger: Learn What You Can Do to Avoid Beach Death Traps," http://www.webmd.com/baby /features/rip-current-no-1-beach-danger.

CHAPTER 7: LANDSLIDES AND MUDSLIDES

1. Name changed for privacy.
2. Author interview with Nicole James, March 2011.
3. Author interview with Bill Harbison, March 2011.
4. Jia-Rui Chong, "Survivor Thanks Her La Conchita Hero," *Los Angeles Times,* January 17, 2005, http://articles.latimes.com/2005/jan/17/local /me-laconchita17.
5. Author interview with Dr. Randall Jibson, United States Geological Survey, March 2011.
6. Author interview with Dr. Jonathan Stock, United States Geological Survey, March 2011.
7. Author interview with Bob Garcia, Sonoma County Animal Control.

CHAPTER 8: TSUNAMIS

1. Author Interview with Dr. Kurt Frankel, March 2011.
2. Ibid.
3. Alison Chung, "Tsunami Warnings After Deadly Chile Quake," Sky News Online, February 28, 2010, http://news.sky.com/home/world-news /article/15561008.
4. Jeff Barnard, "Man Swept into Pacific Sought New Beginning," *Durango Herald,* March 13, 2011, http://www.durangoherald.com/article/20110314 /NEWS03/703149966/-1/news&source=RSS.
5. Author interview with Jenifer Rhoades.
6. Ibid.
7. "Red Cross Emergency Preparedness: Tsunami Education & Awareness," "http://www.rainier-redcross.org/new_web/programs/disaster/Tsunami EducationAwareness.htm.
8. Stephen A. Nelson, "Tsunami," http://www.tulane.edu/~sanelson/geol204/ tsunami.htm.
9. Author interview with Dick Green.

CHAPTER 9: SEVERE THUNDERSTORMS

1. Author interview with Todd Archer, July 2009.
2. Ibid.
3. National Lightning Safety Institute, http://www.lightningsafety.com/.
4. Author interview with Steve Chamberlin, March 2011.

5. National Lightning Safety Institute.
6. "Visitor Precautions," http://www.ultimateyellowstonepark.com/Yellow stone/visitorprecautions.html.
7. "Indoor/Outdoor Swimming Pool Safety," http://www.lightningsafety.com /nlsi_pls/swimming_pools.html.
8. Author interview with Chuck Doswell, April 2011.
9. Author interview with Donna Franklin, March 2011.
10. "Thunderstorm Safety Checklist," http://www.redcross.org/www-files /Documents/pdf/Preparedness/checklists/Thunderstorm.pdf.
11. Author interview with Joanne Yohannan, February 2011.

CHAPTER 10: SNOW AND ICE STORMS

1. Author interview with Shaun Gholston, February 2011.
2. Author interview with Chris Strong, February 2011.
3. "Winter Weather Safety and Awareness," http://www.weather.gov/om /winter/.
4. Ibid.
5. Author interview with Bruce MacInnes, February 2011.
6. Author interview with Staff Sergeant Mark Dornford.
7. Author interview with Joanne Yohannan, February 2011.
8. "Cold Weather Tips," ASPCA, http://www.aspca.org/pet-care/pet-care -tips/cold-weather-tips.aspx.

CHAPTER 11: EARTHQUAKES

1. Inigo Gilmore, "Haiti Earthquake: Survivor Tells of 11 Day Burial Ordeal," *Telegraph*, January 23, 2010.
2. Simon Romero, "Haitian Mother Didn't Give up on Son, Who Eventually Was Pulled From Ruins," *New York Times*, January 23, 2010; Harriet Alexander, "Haiti Earthquake: Two More Survivors in the Rubble as Rescue Scaled Back," *Telegraph*, January 23, 2010.
3. Romero, "Haitian Mother Didn't Give Up on Son."
4. Gilmore, "Haiti Earthquake."
5. Ibid.
6. Ibid.
7. Ibid.
8. Romero, "Haitian Mother Didn't Give Up on Son."
9. Gilmore, "Haiti Earthquake."
10. Author interview Dr. Kurt Frankel, March 2011.
11. South Carolina Department of Natural Resources, http://dnr.sc.gov.
12. "Questions and Answers," earthquakecountryinfo.com, http://www.earth quakecountry.info/faq.php.
13. "Can Animals Predict Earthquakes?" The Examiner.com, http://www .examiner.com/enivronmental-news-in-san-francisco/can-animals-predict -earthquakes.
14. Ibid.
15. "Disaster Planning," At Home Petting Sitting.com, http://www.athomepet sittingsf.com/Pet_Disaster_Plan.html.
16. "National Zoo Animals React to the Earthquake," Smithsonian Biology Conservation Institute, August 24, 2011, http://nationalzoo.si.edu/SCBI/ AnimalCare/News/earthquake.cfm.

CHAPTER 12: HOW TO USE SOCIAL MEDIA IN NATURAL DISASTERS

1. Joplin Tornado Debris Lost and Found Facebook page, https://www.facebook.com/pages/Joplin-Tornado-Debris-Lost-and-Found/223745274319125.
2. Harry Wallop, "Japan Earthquake: How Twitter and Facebook Helped," *Telegraph,* March 13, 2011.19 extreme weather_notes.docx
3. Author Interview with Andew Noyes, March 2011.
4. Author Interview with Sree Sreenivasan, March 2011.
5. Author Interview with Gary Rabkin, March 2011.
6. Author Interview with Gloria Huang.

CHAPTER 13: HOW TO CREATE A FAMILY DISASTER PLAN FOR PEOPLE AND PETS

1. Author interview William Craig Fugate, March 2011.
2. NOAA Weather Radio, http://www.nws.noaa.gov/nwr/.

CHAPTER 14: EMERGENCY SUPPLIES FOR PEOPLE

1. "Are You Ready? Assemble a Disaster Supplies Kit," FEMA website, http://www.fema.gov/areyouready/assemble_disaster_supplies_kit.shtm.
2. "Three Steps to Preparedness," Red Cross website, http://www.redcross.org/portal/site/en/menuitem.d229a5f06620c6052b1ecfbf43181aa0/?vgnextoid=354c2aebdaadb110VgnVCM10000089f0870aRCRD.
3. "How to Create Your Own Roadside Emergency Kit" Edmunds.com, http://www.edmunds.com/how-to/how-to-create-your-own-roadside-emergency-kit.html.
4. Author interview with Staff Sergeant Mark Dornford, February 2011.

CHAPTER 15: EMERGENCY SUPPLIES FOR PETS

1. "Preparedness Fast Facts: Pet Safety," RedCross.org, http://www.redcross.org/portal/site/en/menuitem.53fabf6cc033f17a2b1ecfbf43181aa0/?vgnextoid=a2fb94eeef052210VgnVCM10000089f0870aRCRD&currPage=94ea94eeef052210VgnVCM10000089f0870aRCRD.
2. "Disaster Preparedness," ASPCA.org, http://www.aspca.org/pet-care/disaster-preparedness/; "Make a Disaster Plan for Your Pets," Humane Society.org, http://www.humanesociety.org/issues/animal_rescue/tips/pet_disaster_plan.html#Disaster_Supply_Checklist.
3. "Disaster Preparedness," ASPCA.org.
4. Author interviews with Dr. Charlotte Krugler and Dr. Adam Eichelberger.

INDEX